Mathematics and Mathematicians

Volume 2

Mathematics and Mathematicians

Volume 2

Methods and Problems

P. Dedron and J. Itard

Translated from the French by J. V. Field

The Open University Press
In association with Richard Sadler Ltd

The Open University Press
12 Cofferidge Close, Stony Stratford,
Milton Keynes, MK11 1BY, England

First published 1959 by Editions Magnard, Paris under
the title *Mathématiques et Mathématiciens*.

First published in the English language 1974 by
Transworld Publishers Ltd., in association with Richard
Sadler Ltd.

Published in this edition by The Open University Press 1978
in association with Richard Sadler Ltd.

Made and printed in Great Britain

British Library Cataloguing in Publication Data

Dedron, Pierre
Mathematics and mathematicians.
2. – (Open University. Set books).
1. Mathematics – History
I. Title   II. Itard, Jean   III. Field, Judith
IV. Series
510′.9   QA21

ISBN 0 335 00247 1

Contents of Volume 2

Foreword

By Graham Flegg
Faculty of Mathematics, the Open University

At the beginning of 1973 I was appointed Course Team Chairman for the Open University's second level course, AM 289 *History of Mathematics*, due for first presentation in 1975. One of the first tasks to which the Course Team had to direct its attention was the selection of one or more suitable books which could be used as 'prescribed texts' around which the course itself could be designed. This was by no means an easy task. After studying a not insubstantial number of books, already available in English, I was led to the conclusion that none of these on its own met all the criteria upon which the Course Team's judgement needed to be based. These criteria can be broadly summarized as follows:

(a) The standard of mathematical and historical scholarship should be impeccable.
(b) The content should be intellectually accessible to readers with only a 'general school' level of mathematical background.
(c) The mathematical material should be presented in a general historical context.
(d) There should be a suitable balance between 'mathematics' and 'mathematicians'.
(e) The reader should be exposed to suitable primary source material.

(f) There should be some analysis in depth of suitable specific topics, such as 'written numbers', 'arithmetical calculations', 'solutions of equations', etc.

To find a book meeting all these criteria began to seem well-nigh impossible until the existence of *Mathematique et Mathematiciens* in the original French edition was drawn to my attention by Dr. D. T. Whiteside of Cambridge University. For this, I must record my sincere gratitude to him, and also to Judith Field for undertaking the lengthy task of translation and to Richard Sadler for agreeing to include the English version in the Transworld Student Library. The two-volume work now presented thus results from an original piece of very wise counsel and a substantial amount of very dedicated labour without which it is highly likely that this excellent work would never have been made available to the general English reader. The practising research mathematician of today may perhaps argue that by far the greater part of contemporary mathematics has been invented in the last hundred years and that to present a history of mathematics which excludes the more recent developments is to present only a small fraction of the history of the subject. However, it can also be argued that there is a real sense in which 'history repeats itself' and that to understand the way in which mathematical concepts develop it is not by any means essential to study the mathematics of recent years. Indeed, it may be argued further that a deeper and more balanced historical perspective can be obtained when the material under discussion does not include the contemporary or near-contemporary. This work presents the facts as far as they are known and within the constraints set by its original authors with clarity and authority. It does not attempt to interpret the facts in accordance with any particular view of history in general or of the history of mathematics in particular.* It does not progress far into the nineteenth century, and hence it satisfies the criterion that its content should be accessible to the non-mathematician. However, because the Authors wrote within this constraint, there is no detailed discussion of the invention or development of the infinitesimal calculus, and for this the reader interested in the technical matters associated with the calculus must look elsewhere.† Nevertheless, despite the absence of reference to the more recent developments in mathematics and to the technicalities of the mathematics of Newton and Leibniz, there is a more than adequate wealth of material in these two volumes to satisfy not only the enquiring general reader but also the serious student of the history of mathematics seeking a basis of sound scholarship upon which further and more specialized study can be built. Perhaps above all else, the Authors should be congratulated in introducing the reader to a substantial amount of primary source material which gives some fascinating insight into the minds of great mathematicians of the past, and without which no historical study of mathematics can be seriously undertaken.

MILTON KEYNES
1974

---

* For a stimulating interpretation of the facts relating to the history of number and of geometry the reader is referred to *The Evolution of Mathematical Concepts*—R. L. Wilder—Transworld 1974 (also a prescribed text for the Open University History of Mathematics Course).

† Two particularly useful books are: *The Origins of the Infinitesimal Calculus*—M. E. Baron—Pergamon Press, and *The History of the Calculus and its Conceptual Development*—C. B. Boyer—Dover.

PLATE I. An Egyptian tablet showing numeral hieroglyphs.

1. Written Numbers and Numerical Calculations

Many systems of writing numerals are known, but we shall examine only the written forms, but not the names, of a few of the more important ones.

We shall begin by considering Egyptian numerals and the types of calculation in which they were employed, then go on to describe the sophisticated numerals used by the Sumerians and the Babylonians, then turn to Greek

a unit is represented by:	
ten by:	
a hundred by:	
a thousand by:	
ten thousand by:	
a hundred thousand by:	

---

* So with these nine figures, and with the sign 0 which in Arabic is called Zero, we may write any number we please.

numerals, and finally we shall consider the positional decimal system used by the Hindus and later adopted by the Arabs and the scholars of Europe. We shall trace the development of this system as far as the eighteenth century.

1.1. Egypt

Several units of each order (Fig. 1.1) are represented by repeating the appropriate symbol (See also Plate I).

Numbers can be written from left to right or from right to left, though in the latter case the symbols themselves are turned round.

So 541 can be written either as

 or as

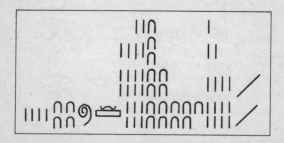.* These numerals are

convenient for carrying out addition and subtraction. The process of multiplication was based on repeated

Fig. 1.2

---

* From hieroglyphic, to hieratic and demotic, the symbols themselves change and the repetitions are replaced by conventional signs.

doubling: for instance in the Rhind papyrus* there is a calculation of the square of twelve, shown in Fig. 1.2, which reads:

1	12	
2	24	
/4	48	
/8	96	sum 144.

In effect, the multiplier is broken down into a sum of powers of two. In this particular case we have $12 = 4 + 8$.

The multiplication is carried out by successive doublings, which give twice, four times and eight times the multiplicand; the products by four and by eight are then added together.

To multiply A by 17 we first note that $17 = 16 + 1$, then calculate $16A$ by four doublings and then add A to our result.

The technique is general.† It has recently come back into use in large computers.

A similar technique was used for division. For instance, to divide 329 by 12: twelve was first doubled repeatedly:

/1	12
/2	24
4	48
/8	96
/16	192
32	384;

then $192 = 12 \times 16$ was subtracted from 329, leaving 137. $96 = 12 \times 8$ was subtracted from 137, leaving 41. Subtracting $24 = 12 \times 2$ left 17, and finally subtracting $12 = 12 \times 1$ left 5. The quotient is given as $16 + 8 + 2 + 1 = 27$.

---

*The Rhind papyrus is in hieratic. The example we give here has been rewritten in hieroglyphic.

†In 1617 Napier used this same method to carry out a rapid calculation with counters, which he describes in Arithmetica Localis, the last part of his *Rhabdologia*.

To multiply by 10 or by 100 was a simple matter and the Egyptians in fact used such multiplications to shorten calculations. For example, in one of the calculations in the Rhind papyrus we are required to divide 1120 by 80. Now, $80 \times 10 = 800$, that is

 multiplied by 10 is written

The calculation reads

1	80
/10	800
2	160
/4	320
Total	1120

The quotient is 14.

The Egyptian method of handling fractions is important, since it was adopted by the Greeks, who used it for a considerable time, alongside more sophisticated methods.

The fractions used by the Egyptians (apart from a few exceptions, such as $\frac{2}{3}$ and $\frac{3}{4}$) all have the numerator 1. That is, they are the reciprocals of integers (i.e. integral parts). Each part is denoted by its denominator with a sign ⬭ written above it.

So $\frac{1}{15}$ is written ⬭∩∣∣∣ . There is a special sign for

$\frac{2}{3}$, namely ⬭∏ .

We shall adopt similar conventions and write $\frac{2}{3}$ as $\overline{\overline{3}}$, $\frac{1}{15}$ as $\overline{15}$, $\frac{1}{3}$ as $\overline{3}$ etc. . . .

Addition is indicated by juxtaposition: $\frac{1}{4} + \frac{1}{5}$ is written $\overline{4}\ \overline{5}$.

However, frequent use is made of identities such as

$$\bar{6}\ \bar{6} \text{ equals } \bar{3}$$
$$\bar{6}\ \bar{6}\ \bar{6} \text{ equals } \bar{2}$$

$$\bar{3}\ \bar{3} \text{ equals } \bar{\bar{3}}$$
$$\bar{3}\ \bar{6} \text{ equals } \bar{2}$$
$$\bar{2}\ \bar{3}\ \bar{6} \text{ equals } 1$$

$$\bar{\bar{3}} \text{ equals } \bar{2}\ \bar{6}$$

$$\bar{2}\ \bar{3} \text{ equals } \bar{\bar{3}}\ \bar{6}$$

$$\bar{\bar{3}}\ \bar{2} \text{ equals } 1\ \bar{6}$$

The basic arithmetical operation is that of doubling.

The Rhind papyrus contains a table of the results obtained by doubling all the fractions with odd denominators, as far as the fraction with denominator 101. We show part of this table on p. 14. The first column contains the denominator of the fraction which is to be doubled, and the second the denominators of the parts whose sum is the required result.

For example, the number 21 in the first column corresponds to 14 and 42, which means that $\frac{2}{21} = \frac{1}{14} + \frac{1}{42}$.

1.2. Mesopotamia

The sophisticated system of numerals invented by the Sumerians is completely different from the crude system which was normally employed by the Babylonians.

Except for the modern decimal system it is the much older Sumerian sexagesimal system that has exerted most influence on the progress of science, although all that now remains of it is that angles and intervals of time are still divided into sixtieths.

The Sumerian system uses only two symbols: ｜ for units and ＜ for tens.

5	3	15		
7	4	28		
9	6	18		
11	6	66		
13	8	52	104	
15	10	30		
17	12	51	68	
19	12	76	114	
21	14	42		
23	12	276		
25	15	75		
. . . . . .				
81	54	162		
83	60	332	415	498
85	51	255		
87	58	174		
89	60	356	534	890
91	70	130		
93	62	186		
95	60	380	570	
97	56	679	776	
99	66	198		
101	101	202	303	606

Numbers up to 59 are written in the Egyptian manner: for example

34:

47:

However, 60 is written exactly like 1. Hence, for example:

61: **❚ ❚** (60+1)

92: **❚ ❰ ❰ ❰ ❚❚** (60+32)

The Old Babylonian texts do not use a symbol for zero: missing units are merely indicated by a space. Thus 3601 = 60 × 60 + 1 is written **❚ ❚**.

Numbers written in this way, without a symbol for zero, can sometimes be difficult to read, but since the base 60 is so large problems do not arise as often as they would in a system with a smaller base such as 10.

The Babylonian system also uses position to indicate negative powers of sixty exactly as the modern decimal system uses position to indicate negative powers of ten.

In the decimal system we can write an approximate value of $\sqrt{2}$ in the form 1·41421.

The Babylonians write

i.e.

$$1\ 24\ 51\ 10 \text{ or } 1 + \frac{24}{60} + \frac{51}{3\ 600} + \frac{10}{216\ 000}.$$

Apart from using a different base, the Babylonian system differs from the modern one in that it does not mark the position of the units explicitly: there is no equivalent to our decimal point. We could therefore say that the Babylonian system employed a "floating" point.

In what follows we shall write Babylonian numbers in Arabic numerals but using base 60, as in the expression for $\sqrt{2}$ given above.*

These sexagesimal numbers were by far the best arithmetical system of ancient times.

Addition and subtraction were carried out exactly as we should carry them out now if we were working in hours, minutes and seconds.

Multiplication, however, involved the use of tables much more sophisticated than our own. Archaeologists have found a considerable number of such tables, written with a stylus on clay tablets about the size of a man's hand and three or four centimetres thick.

We might at first expect that the multiplication tables would go from 1 to 59, but they do not in fact do so. The explanation lies in the technique used for carrying out division.

Nowadays schoolchildren are taught that to multiply or divide a number by 10, 100 or 1000 the decimal point is moved one, two or three places to the right or to the left; to multiply a number by 0·5 it is divided by 2; to divide a number by 5 it is doubled, and the decimal point is moved one place to the left, etc. ...

Short cuts like these are possible because 5 and 2 are factors of the base, 10.

Sixty, the base of the Babylonian system, has many more factors, namely: 60, 30, 20, 15, 12, 10, 6, 5, 4, 3 and 2.

Thus, many numbers which can be expressed in one, two or three sexagesimal figures have reciprocals which can

---

* For ease of reading, we shall indicate the order of the various units when we state and solve problems. The principal unit will be followed by a semicolon (;) and the others by a comma (,).

also be expressed in this way. Such numbers will be described as *regular*.

In a system with base 10 the reciprocal of an integer can be expressed exactly as a decimal fraction if the integer is of the form $2^a \times 5^b$.

In a sexagesimal system, the reciprocal of any number can be expressed exactly if and only if the number is of the form $2^a \times 3^b \times 5^c$.

The first integer which does not satisfy this condition is 7. The Babylonians expressed this fact by saying that seven does not divide.

We show on p. 18 a transcription of one of the many Babylonian tables of reciprocals that have been found. It reads as follows:

$$\frac{1}{2} = \frac{30}{60} \cdots \frac{1}{8} = \frac{7}{60} + \frac{30}{60 \times 60};$$

$$\frac{1}{9} = \frac{6}{60} + \frac{40}{60 \times 60} \quad \text{etc.}$$

Multiplication tables give as multipliers the numbers 1, 2, 3, 4 ... up to 20, and then the numbers 30, 40 and 50. All the multiplicands are regular numbers, that is, they have exact reciprocals. The table therefore serves two purposes: it can be used for multiplication and also for division. To divide 15 by 16 we first look up 16 in a table of reciprocals, which gives us 3,45; we then look up 3,45 in the multiplication table, which gives us 3,45 × 15 = 56,15.

Tables of squares, cubes, etc., have also been found.

We know that the Babylonians knew how to extract square roots since, as can be seen from the example we gave above, they were familiar with a value of $\sqrt{2}$. They must have found such roots by a method something like the modern one. We shall return to this question later.

Number	Reciprocal	Number	Reciprocal
2	30	27	2, 13, 20
3	20	30	2
4	15	32	1, 52, 30
5	12	36	1, 40
6	10	40	1, 30
8	7, 30	45	1, 20
9	6, 40	48	1, 15
10	6	50	1, 12
12	5	54	1, 6, 40
15	4	1	1
16	3, 45	1, 4	56, 15
18	3, 20	1, 12	50
20	3	1, 15	48
24	2, 30	1, 20	45
25	2, 24	1, 21	44, 26, 40

The mathematical and astronomical tablets from the Seleucid era (that is, after the time of Alexander the Great) employ a numerical system which includes a symbol for zero. The zero occurs either at the beginning of a number or within it, but never at the end.

1.3. Greece

The Greeks employed various systems of numerals but we shall be concerned only with the sophisticated system which seems to date from the fifth century BC.* It was used by the Jews, and then by the Arabs, and by all European mathematicians until the modern decimal system was introduced. Even after this, Arab astronomers continued for some time to use the old Greek system. It was therefore in use throughout the Mediterranean area for more than a millenium.

The Greek system, unlike the Babylonian, does not use strictly positional notation.

---

* It was not, however, adopted as the official system in Athens until the first century AD. H.G.F.

Simple units are denoted in order by

$$\alpha \ \beta \ \gamma \ \delta \ \epsilon \ \digamma \ \zeta \ \eta \ \theta$$

that is, by the first nine letters of the Greek alphabet, the number 6 being written as ϛ (digamma), a letter not found in the alphabet of the time.

Tens are expressed by nine further letters:

$$\iota = 10, \kappa = 20, \lambda, \mu, \nu, \xi, o, \pi, \ \rho = 90$$

(ϟ, koppa, is, like ϛ, digamma, a letter not used in the ordinary alphabet).

The process is repeated for hundreds:

$$\rho = 100, \sigma = 200, \tau, \upsilon, \phi, \chi, \psi, \omega, \text{ϡ} = 900.$$

Here again 900 is represented by an archaic letter ϡ, sampi.

For thousands the same letters are used as for the corresponding units, but an accent is added below them to their left:

$$,\alpha = 1000, \quad ,\beta = 2000\ldots, \quad ,\digamma = 6000\ldots, \quad ,\theta = 9000.$$

Myriads (tens of thousands) are often denoted by writing the number of myriads above a letter M: *e.g.* $\overset{\beta}{M} = 20{,}000$.

So, 248 is written $\overline{\sigma\mu\eta}$, the bar being added to avoid confusing number with words.

Also, 5630 is written as $,\epsilon\chi\lambda$.

As can be seen from this example there is no need for a symbol for zero in the Greek system of numerals.

Aristarchus of Samos writes 7175,5875 as $\overset{,\zeta\rho o\epsilon}{M} ,\epsilon\omega o\epsilon$.

The system changed over the centuries. Diophantus, for example, writes the number of myriads to the right of the myriad sign $\overset{\curlyvee}{M}$ and separates it from the rest by a dot:

$$150{,}7984 \text{ is written } \overset{\curlyvee}{M}\rho\nu.,\zeta \text{ϡ}\pi\delta.$$

Archimedes proposed that for writing very large numbers a kind of positional notation should be used. The numbers were to be expressed in terms of powers of 10^8 (*octades*) and then in powers of $10^{8.10^8}$.

This system was proposed as purely theoretical and it was never actually used.

Pappus has left us a partial account of the system used by Apollonius. It employs successive powers of a myriad :

$$\mu^{\gamma} \ ,\varepsilon\upsilon\xi\beta \ \ \kappa\alpha\grave{\imath} \ \ \mu^{\beta} \ ,\gamma\chi \ \ \kappa\alpha\grave{\imath} \ \ \mu^{\alpha} \ ,\varsigma\upsilon$$

meaning: third powers of a myriad 5462; and (καὶ) second powers of a myriad 3600; and myriads 6400, which adds up to 5 462 360 064 000 000.

A slightly modified form of Apollonius's system was used as late as the fourteenth century by the Byzantine mathematician Nicolas Rhabdas.

The Jews and the Arabs also used the Greek system of numerals, though they adapted it to suit their own alphabets.

In the Greek system arithmetic operations involving integers are carried out in more or less the same way as they are in the modern system, though they are rather more complicated.

For example, to add two decadic numbers κ and μ we must first read them and identify them as two tens and four tens. They are then added up mentally, as now, to give six tens. This is written down as ξ. In modern positional notation the first and last parts of the calculation can be done instantaneously and automatically. In the same way, when multiplying λ by χ a Greek, if he did not use a multiplication table, would first have to read the two numbers as three tens and six hundreds.

Multiplying tens by hundreds gives thousands.

Like us, a Greek would have worked out in his head that six times three are eighteen. Eighteen thousand, i.e. one myriad and eight thousand, is written: $\overset{\alpha}{\text{M}},\eta$.

Greek methods of calculation were much less economical than modern ones. Tables were therefore much more necessary to the Greeks than they are to us. Nicolas Rhabdas gives very elaborate tables for addition and multiplication, though he does also explain briefly how to manage without them.

Figure 1.3 shows an example of multiplication taken from Eutocius's commentary on Archimedes's treatise *On the Measurement of the Circle*.

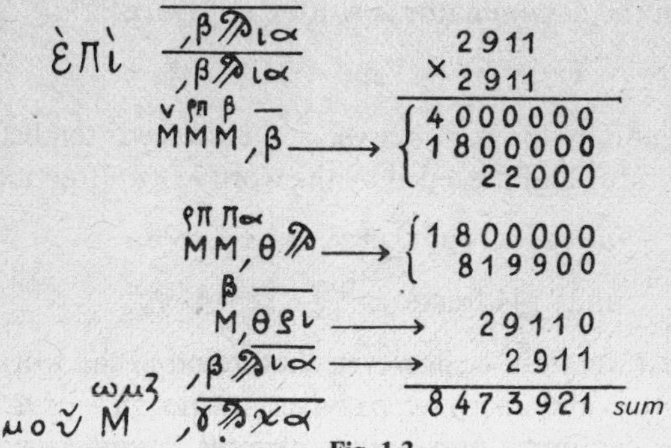

Fig. 1.3

The Greeks handled fractions in three different ways: they used the reciprocals of integers, like the Egyptians, and also general fractions, while in astronomical work they used sexagesimal fractions, derived from the Babylonians.

A reciprocal is generally denoted by its denominator with an accent written above it to the right:

$\frac{1}{3}$ is written γ', $\frac{1}{32}$ $\lambda\beta'$ and $\frac{1}{112}$ $\rho\iota\beta'$.

There are, however, special signs for $\frac{1}{2}$: V' or C' and for $\frac{2}{3}$: w'.

A number which contains a fraction is written as follows:

$\kappa\theta w'\iota\gamma'\lambda\theta' = 29\frac{2}{3}\frac{1}{13}\frac{1}{39}$ i.e. $29\frac{10}{13}$

$\mu\theta V'\iota\zeta'\lambda\delta'\nu\alpha' = 49\frac{1}{2}\frac{1}{17}\frac{1}{34}\frac{1}{51}$ i.e. $49\frac{31}{51}$.

A part of a part is written as:

$\iota\gamma'$ τοῦ $\iota\gamma'$ meaning $\frac{1}{13}\times\frac{1}{13}=\frac{1}{169}$.

Sometimes there are two accents instead of one:

$$ζ''\quad\text{for}\quad\tfrac{1}{7}.$$

When general fractions are used (i.e. fractions which may have any number as numerator) they are written in various ways. Archimedes writes $\frac{10}{71}$ as ι οα′, the numerator, without an accent, preceding the denominator, which is accented.* He also writes $1838\frac{9}{11}$ as ͵αωλη θ ια′. This notation can lead to confusion, and we find, for example, that in Heron's collection the denominator is written out twice:

$$ειγ′ιγ′ = \tfrac{5}{13}\quad\text{and}\quad ⊏ζ′ζ′ = \tfrac{6}{7}.$$

The denominator is, moreover, sometimes written before the numerator and preceded by the word λεπτα (fractions):

$$μονάδες\ ρμδ\ λεπτὰ\ ιγ′\ ιγ′\ σϙθ$$

units 144 fractions $\frac{299}{13}$ i.e. $144\frac{299}{13}$.

The best notation is, however, that found in the work of Diophantus, and in some passages of Heron's *Metrica*, where the denominator is written *above* the numerator:

$$\frac{ιⵉ}{ρκα}\quad\text{for}\quad\frac{121}{16};\qquad\frac{ρκη}{ρ}\quad\text{for}\quad\frac{100}{128}$$

and

$$\frac{βνβ}{τπθV′}\quad\text{for}\quad\frac{389\frac{1}{2}}{152}.$$

Occasionally the denominator is written above the numerator to the right, like an exponent:

$$\overline{ιε}δ\quad\text{for}\quad\frac{15}{4}.$$

---

* The reader should note that since we know very little about the conventions which governed writing in Classical times we cannot maintain with any degree of certainty that the notation which has come down to us is identical with that used by the original authors.

After the second century BC all Greek astonomers used sexagesimal fractions. Their system was based on the Babylonian one, but it applied only to negative powers of 60, that is, to fractions. One weakness of the system was that the Greek decimal notation was employed for integers. A zero was used, and unlike the zero found in third-century Babylonian work it was used either at the beginning of a number, embedded in it, or at the end. Ptolemy writes zero as ō.* This cannot be confused with o = 70 since decadic numbers are used only up to 50.

For example in Book III of Ptolemy's *Almagest* we find a table giving the mean motion of the sun:

'Οκτὼ καὶ δεκαιτῶν	Μοιρ	A	B	Γ	Δ	E	F
λϛ	τνα	ιδ	να	ιβ	μα	θ	ō
χλ	σϛ	μθ	νϛ	ιβ	ō	ζ	λ

That is

Periods of 18 years	Degrees	Min	Sec	Thirds	Fourths	Fifths	Sixths
36	351	14	51	12	41	9	0
630	206	49	56	12	0	7	30

1.4. Latins

We shall refer to the scholars of the Latin West not for their learning but merely to remind ourselves of their ideas so that we can appreciate the importance of the advances brought about by their contacts with the Arabs, from about 1000 AD onwards.

The fall of the Western Empire cut off the Latins from the Eastern Mediterranean. They continued to use Roman

* Various other symbols had been used in Ptolemaic papyri:

numerals, which are in fact still employed to this day for certain purposes. Roman numerals somewhat resemble an archaic Greek system, which we have not discussed, and are now written in the form:

1	2	3	4	5	6	7	8	9	10	50	100	500	1000
I	II	III	IV	V	VI	VII	VIII	IX	X	L	C	D	M

While it is not absolutely impossible to carry out calculations with Roman numerals, it is not easy to do so, and the numerals are in fact more cumbersome than the more sophisticated Greek numerals.

The Roman system of writing and manipulating fractions is now less well-known since it is no longer used.

The system of seventeen fractions described by Victor of Aquitaine in the fifth century AD is shown in the table opposite (Fig. 1.4).

A more elaborate system is sometimes found, for example in Campanus's Euclid.

The first column gives the name of the fraction, the second the symbol used for it by Victor of Aquitaine (as edited by G. Friedlein in 1871) and the third column gives the symbol used in the edition of Campanus's Euclid* published in Basle in 1558.

Roman fractions are both complicated and inadequate. They require even more translation than the integers do before calculations can be performed with them, and they are far more cumbersome than the Greek fractions.

It is understandable that in such circumstances the abacus† and numerical tables were used as aids to calculation. Figure 1.5 shows part of Victor of Aquitaine's multiplication tables, in each of which the value of the

---

* Campanus lived in the thirteenth century. The first printed edition of his Euclid was published in Venice in 1482 (see Vol. I, p. 187). J.V.F.

† See below p. 27*ff*.

NAME	Victor of Aquitaine	Campanus	VALUE	
			As =1	Uncia =1
As	1	p	1	12
Deunx	SSS·	S=-=	11:12	11
Dextans	SSS	S==	5:6	10
Dodrans	SS·	S ʒ	3:4	9
Bes	SS	-S-	2:3	8
Septunx	S˧	√‾	7:12	7
Semis	S	S	1:2	6
Quincunx	ʒſ˧	=-=	5:12	5
Triens	ʒſ	= =	1:3	4
Quadrans	ʒ˧	=-	1:4	3
Sextans	ʒ	=	1:6	2
Sexuncia	ꝡ		1:8	3:2
Uncia	⁄	—	1:12	1
Semuncia	ꝇ	Ɛ	1:24	1:2
Duella	UU	U U	1:36	1:3
Sicilicus	Ɔ	⊙‾	1:48	1:4
Sextula	U	U	1:72	1:6
Drachma		⊬	1:96	1:8
Dimidio sextula	4	c|ɔ	1:144	1:12
Tremissis		H	1:216	1:18
Scrupulus		⅄⅄	1:288	1:24
Obulus		ᴄ⁊	1:576	1:48
Bissiliqua		'M'	1:864	1:72
Cerates		Z	1:1152	1:96
Siliqua		C‖I	1:1728	1:144
Chalcus		Ǫ	1:2304	1:192

Fig. 1.4

multiplicand runs from a thousand units to a dimidio sextula, i.e. $\frac{1}{144}$.

The multiplier is constant for each table, and takes values from 1 to 50. Our examples show multiplication by 10 and by 48.

We note that the thousands are written not as M but as $\overline{1}$.

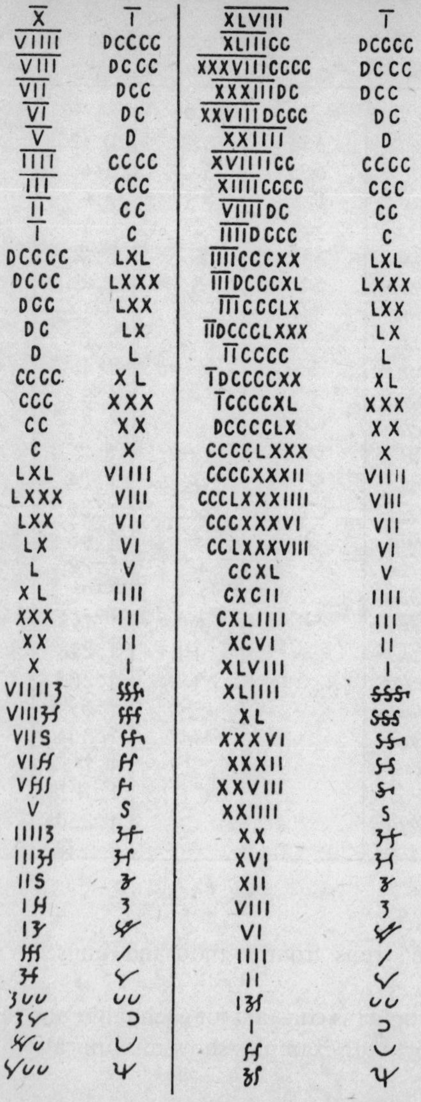

X̄	T̄	X̄LVIII	T̄
V̄IIII	DCCCC	X̄LIIICC	DCCCC
V̄III	DCCC	XXXVIIICCCC	DCCC
V̄II	DCC	XXXIIIDC	DCC
V̄I	DC	XXVIIIDCCC	DC
V̄	D	XXIIII	D
ĪĪĪĪ	CCCC	XVIIIICC	CCCC
ĪĪĪ	CCC	XIIIICCCC	CCC
ĪĪ	CC	V̄IIIIDC	CC
Ī	C	IIIIDCCC	C
DCCCC	LXL	ĪĪĪĪCCCXX	LXL
DCCC	LXXX	ĪĪĪDCCCXL	LXXX
DCC	LXX	ĪĪCCCLX	LXX
DC	LX	ĪĪDCCCLXXX	LX
D	L	ĪĪCCCC	L
CCCC	XL	T̄DCCCCXX	XL
CCC	XXX	T̄CCCCXL	XXX
CC	XX	DCCCCLX	XX
C	X	CCCCLXXX	X
LXL	VIIII	CCCCXXXII	VIIII
LXXX	VIII	CCCLXXXIIII	VIII
LXX	VII	CCCXXXVI	VII
LX	VI	CCLXXXVIII	VI
L	V	CCXL	V
XL	IIII	CXCII	IIII
XXX	III	CXLIIII	III
XX	II	XCVI	II
X	I	XLVIII	I
VIIII3	SSS	XLIIII	SSS
VIII3	SSS	XL	SSS
VIIS	SS	XXXVI	SS
VISS	SS	XXXII	SS
VSII	S	XXVIII	S
V	S	XXIIII	S
IIII3	SS	XX	SS
IIIS	SS	XVI	SS
IIS	S	XII	S
ISII	S	VIII	S
IS	S	VI	S
SSS	/	IIII	/
SS	Y	II	Y
3UU	UU	13I	UU
3S	U	I	U
SUU	U	SS	U
SUU	4	3S	4

Fig. 1.5

It can also be seen that since Roman fractions do not follow a decimal progression multiplying by 10 is less simple than for integers. On the other hand, multiplying the fractions by 48 gives very simple results.

1.5. Hindus, Arabs and Western Europe

In the third century BC, during the reign of the Buddhist King Asoka, two different systems of numerals were in use among Hindu mathematicians. One, Kharosti, was rather like the primitive Greek or Latin system and the other, Brahmi, very like the more sophisticated Greek system.

It was the sixth or seventh century AD that the system of numerals that we use today, positional decimal arithmetic including a zero first appeared in India. The only difference between the Indian system and the modern one is in the signs used for the figures. These signs have in fact taken very different forms at various times and in various places, and in Western Europe their form became fixed only with the introduction of printing.

The Arabs adopted the positional system in the ninth century and their conquests introduced it to the West.

As we have pointed out, calculation with Roman numerals was difficult and the Latins therefore made use of counters (*calculus* = a pebble). They employed a board divided up into columns, which corresponded to units, tens, hundreds, etc. . . . . Numbers were represented by identical counters placed in the appropriate columns. For instance, the number MCCCCLVI was represented by six counters in the unit column, five in the tens column, four in the hundreds column and one in the thousands column.

This method made additions and subtractions very easy, but it was less successful for multiplication and, more particularly, for division. There are many sixteenth-, seventeenth- and even eighteenth-century treatises which enable us to form some idea of how such calculations were

B.N.

PLATE II. Two
methods of calculation:
with pen and paper, and
with an abacus. From
Margarita Philosophica
(1504).

made. For instance, Jean Trenchant explains how to
multiply 767 by 46. On one side of his abacus (which is
different from a Roman one) he sets out the counters to
represent the number 767. He then takes away one of these
counters and substitutes for it 46 counters on the other
side of the abacus, placed in the appropriate column. He
continues in this way until he has removed all the counters
of the multiplicand. The procedure is slow. Juan Martinez
Siliceus uses a multiplication table when multiplying with
an abacus. Calculating with an abacus is thus very like
calculating with pen and paper (Plate II). Moreover, by
this time the limitations of the abacus were well-known.
Simon Jacob (who died in Frankfurt in 1564) says:

It is true that it [the abacus] is useful for everyday calculations, where
we often need to calculate sums, subtract or add, but in technical work,
which is rather more complicated, an abacus is often an encumbrance.
I am not claiming that such calculations cannot be carried out on the lines
[of the abacus] but, just as a man without baggage has the advantage over
one who is heavily loaded, so calculating with figures has the advantage
over calculating on the lines.

An improvement was introduced in the tenth century: the identical counters were replaced by counters marked with the numbers $1, 2, 3, \ldots, 9$ (in a form very different from the modern one). The number MCCCCLVI is then represented on the abacus by the counter 6 in the unit column, a 5 in the tens column, a 4 in the hundreds column, and a 1 in the thousands column.

Calculations, which require a knowledge of the multiplication table up to 9×9, are then as easy as they are today, although the technique used for division is surprisingly different from the one we use now.*

In the thirteenth century the abacus (which had possibly been introduced into Europe through the Arabs of Spain) was finally replaced by the *Algorithm*, that is to say, the Hindu method of calculation, using a zero. From then on calculations could be performed 'in writing'.

Although the use of positional decimal notation, *for integers only*, now began to spread throughout the West, calculations were not carried out exactly as they are now. Their present form is the result of a further five centuries of development.

In Lazarus Schöner's edition of Ramus's *Arithmetic*, printed in 1586, subtraction starts with the largest units.

Fig. 1.6

---

* The multiplication table was often learned only up to 5×5. For numbers between 5 and 10 the so-called 'lazy rule' (*regula pigri*) was used. Suppose we want to find 8×6. We subtract both figures from 10, obtaining complements 2 and 4. We take 2 from 6 or 4 from 8. This gives 4, the number of tens in the required product. The product of the complements $2 \times 4 = 8$ gives the number of units.

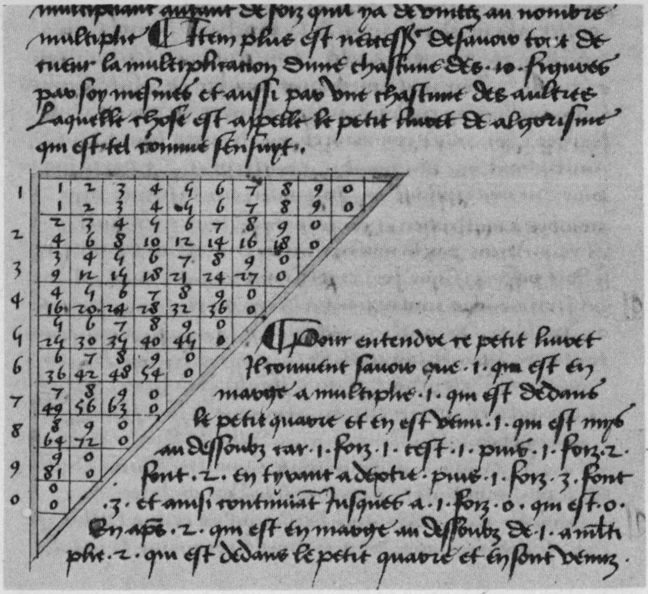

multipliant autant de foiz quil ya de unitez au nombre multiplie. ¶ Item plus est necessaire de savoir tout de cueur la multiplicacion dune chascune des .10. figures par soy mesmes et aussi par une chascune des aultres Laquelle chose est appellee le petit livret de algorisme qui est tel comme sensuyt.

1	1 1	2 2	3 3	4 4	5 5	6 6	7 7	8 8	9 9	0 0
2	2 4	3 6	4 8	5 10	6 12	7 14	8 16	9 18	0 0	
3	3 9	4 12	5 15	6 18	7 21	8 24	9 27	0 0		
4	4 16	5 20	6 24	7 28	8 32	9 36	0 0			
5	5 25	6 30	7 35	8 40	9 45	0 0				
6	6 36	7 42	8 48	9 54	0 0					
7	7 49	8 56	9 63	0 0						
8	8 64	9 72	0 0							
9	9 81	0 0								
0	0 0									

¶ Pour entendre ce petit livret il convient savoir que .1. qui est en marge a multiplie .1. qui est dedans le petit quarre et en est venu .1. qui est mys au dessoubz car .1. foiz .1. est .1. puis .1. foiz .2. font .2. en tirant a dextre. puis .1. foiz .3. font .3. et ainsi continuant jusques a .1. foiz .0. qui est .0. En apres .2. qui est en marge au dessoubz de .1. multiplie .2. qui est dedans le petit quarre et en sont venuz.

B.N.

PLATE III. A page from the *Triparty* (1484) of Nicolas Chuquet.

This very logical procedure is quite convenient when using an abacus or writing on a slate, but when a pen is used it involves an irritating amount of crossing-out. Our Fig. 1.6 is taken from Schöner's book and shows the subtractions 345 – 234 and 432 – 345.

To illustrate multiplication we show a page from the *Triparty* of Chuquet (Plate III).

The curious triangular shape of this table is found again in works dating from the sixteenth century, for instance those of Finé and of Jean Trenchant.

Multiplication soon took on its modern form.

However, Chuquet, Pacioli and Tartaglia mention a method, known as multiplication by parallelograms or by lattices, which, at the time they were writing, was no more than an antiquated survival. It looks like Napier's *Rabdo-*

logia and perhaps originated from a form of calculation with marked counters which antedated written calculation. Tartaglia provides the following example (Fig. 1.7):

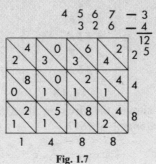

Fig. 1.7

We require to multiply 4567 by 326.

On the right we carry out the "proof by 7".*

Division was the elementary operation which took the longest time to find a stable form. In France, the methods of division which were used up to the beginning of the eighteenth century were as clumsy as Ramus's method of subtraction.

---

* Proofs by 9 and by 7.

Proof by 9 is found in Hindu and Arab work of the tenth century. About the year 1000 al-Karkhi carried out proofs by 9 and by 11. In 1202, Leonardo of Pisa used proof by 9 and mentioned proofs by 7 and 11. In 1484 Nicolas Chuquet used proof by 9, and at about the same time Luca Pacioli used proofs by 9 and by 7. Juan Martinez Siliceus (1514), Tartaglia (1557) and Clavius (1583) explicitly use proofs by 9 and by 7. All three describe the same method of finding the remainder after dividing 15683 by 7: We divide 15 by 7, the remainder is 1. Then 16 by 7, remainder 2. Then 28 by 7, remainder 0. Then 3 by 7, remainder 3.

In 1556 Forcadel describes a modification of this method. He notes that $10 = 7 + 3$, and in the example we have given multiplies 1 by 3 and adds the result to 5, giving 8, whose remainder is 1. He then calculates $1 \times 3 + 6 = 9$, remainder 2. $2 \times 3 + 8 = 14$, remainder 0. Remainder $0 + 3 = 3$.

The two proofs apply to the four operations. In an edition of the work of Juan Martinez Siliceus, published in 1526, the editor notes that one can also carry out proofs by 5. In short, proofs by 9, 11, 7 etc. . . . had been commonplace at least since the tenth century.

Fig. 1.8

Figure 1.8 shows the division of 6754 by 357 from Legendre's *Arithmetic* of 1647.

FRACTIONS

The Hindus used the modern notation for fractions, but without the horizontal line. The Arabs at first copied the Hindu notation but later introduced the horizontal line.

Both the Hindus and the Arabs on the whole handled fractions as we do now. There was, however, one form of calculation which is now no longer used but was once quite important, appearing in the work of the Arabs, and of Leonardo of Pisa, and even in works of sixteenth century: the use of continuous rising fractions, which Leonardo calls "fractiones in gradibus", step fractions. For example, he writes $\cdot \frac{1}{2} \frac{4}{7}$ and explains that this must be read as $\frac{4}{7}$ plus $\frac{1}{2}$ a seventh, i.e. $\frac{4}{7} + \frac{1}{14}$. In the same way $\cdot \frac{1}{2} \frac{0}{7}$ is to be read as zero sevenths plus $\frac{1}{2}$ a seventh, and

$$\cdot \frac{1}{2} \frac{5}{6} \frac{7}{10} \quad \text{is} \quad \frac{7}{10} + \frac{5}{10 \times 6} + \frac{1}{10 \times 6 \times 2}.$$

The fact that Leonardo works from right to left is due to the influence of the Arabs, who write in this direction.

In his *Book of Square Numbers* Leonardo writes $\frac{9}{10} \frac{6}{10} 13$ for the number we should write as 13·69, and in this particular case his notation is very like that of modern decimals.

However, neither the Hindus, nor the Arabs, nor, until the end of the sixteenth century, the scholars of the West, realised how useful it would be to extend positional decimal notation to the right as well as to the left as the Babylonians had done with their sexagesimal notation.

This delay was, in fact, due to the very success of the Babylonian sexagesimal system, which had been adopted by Greek astronomers in the second century BC and was still used by the Arabs and in the West (all astronomical tables had been calculated with this base).* It was used for astronomical calculations until the seventeenth century† and it is still used today for angles and for time.

Signs of a coming reform were, however, to be found in trigonometry and in the calculation of square, cube or higher roots.‡

Vieta writes in his *Universalium inspectionum ad Canonem Mathematicum*, published in 1576 as a supplement to his *Mathematical Canon* (*Canon Mathématique*), that: "In Mathematics sixtieths and sixties should be used rarely or not at all. The opposite applies to thousandths and thousands, hundredths and hundreds, tenths and tens and other pairs of the same kind, larger or smaller: these should be used frequently or exclusively."

When Vieta wants to find two mean proportionals between the radius of a circle and the side of the inscribed triangle he divides the radius into 100 000 parts and writes

---

* In the Alphonsine tables (thirteenth century) a sexagesimal base is even used in both directions, to the left as well as to the right, which is an astonishing survival from Sumerian mathematics.

† In the sixteenth century sexagesimal fractions were called "physical numbers".

‡ See Chapters 3 and 6.

down the following results:

$$100,000,\underline{000} \quad 112,246,\underline{205} \quad 125,992,\underline{106} \quad 141,421,\underline{356}$$

He uses a comma to separate the figures of the number into groups of three. The part expressed as a decimal fraction is written in slightly smaller figures and is underlined. The denominator of 1000 is understood.

In a later passage of the same work Vieta merely separates the fractional part from the integral part with a vertical line.

The last step was taken by Stevin, in 1582, in a work called *De Thiende*, which he himself translated into French in 1585 under the title of *La Disme*. We quote from an English translation by Robert Norton, published in 1608.

THE PREFACE OF SIMON STEVIN

To Astronomers, Land-meters, Measurers of Tapestry, Gaugers,
Stereometers in general, Money-Masters, and to all
Merchants, SIMON STEVIN wishes health.

Seeing then that the matter *of this Dime (the cause of the name whereof shall be declared by the first* definition *following) is number, the use and effects of which yourselves shall sufficiently witness by your continual experiences, therefore it were not necessary to use many words thereof, for the* astrologer *knows that the world is become by* computation astronomical *(seeing it teaches the pilot the elevation of the* equator *and of the* pole, *by means of the declination of the sun, to describe the true longitudes, latitudes, situations and distances of places, etc.) a paradise, abounding in some places with such things as the earth cannot bring forth in other. But as* the sweet is never without the sour, *so the travail in such computations cannot be unto him hidden, namely in the busy multiplications and divisions which proceed of the 60th progression of degrees, minutes, seconds, thirds, etc. And the surveyor or land-meter knows what great benefit the world receives from his science, by which many dissensions and difficulties are avoided which otherwise would arise by reason of the unknown capacity of land; besides, he is not ignorant (especially whose business and employment is great) of the troublesome multiplications of rods, feet, and oftentimes of inches, the one by the other, which not only molests, but also often (though he be very well experienced) causes error, tending to the damage of both parties, as also to the discredit of landmeter or surveyor, and so for the money-masters, merchants, and each one in his business. Therefore how much they are more worthy, and the means to attain them the more laborious, so much the greater and better is this Dime, taking away those difficulties. But how? It teaches (to speak in a word) the easy performance of all reckonings, computations, & accounts, without broken numbers, which can happen in man's business, in such sort as that the four principles of arithmetic, namely addition, subtraction, multiplication, & division, by whole numbers may satisfy these effects, affording the like facility unto those that use counters. Now if by those means we gain the time which is precious, if hereby that be saved which otherwise should be lost, if so the pains, controversy, error, damage, and other inconveniences commonly happening therein be eased, or taken away, then I leave it willingly unto your judgement to be censured; and for that, that some may · say that certain inventions at the first seem good, which when they come to be practised effect nothing of worth, as it often happens to the searchers of strong moving, which seem good in small proofs and models, when in great, or coming to the effect, they are not worth a button: whereto we answer that herein is no such doubt, for experience daily shows the same, namely by the practice of divers expert land-meters of Holland, unto whom we have shown it, who (laying aside that which each of them had, according to his own manner, invented to lessen their pains in their computations) do use the same to their great contentment, and by such fruit as the nature of it witnesses the due effect necessarily follows. The like shall also happen to each of yourselves using the same as they do. Meanwhile live in all felicity.*

Stevin's notation, which is based on algebraic symbols, is not exactly like that used today. For example, Stevin wrote 872.53 in the form 872 ⓪ 5 ① 3 ②.

For addition

					⓪	①	②
345·72					345	7	2
872·53	is written in *La Disme* as				872	5	3
615·48					615	4	8
956·86					956	8	6
2 790·59					2 790	5	9

In 1592 G. A. Magini used a point to separate the integral part of the number from the fractional part: 15·378.

From 1592 onwards Bürgi's manuscript work uses the notation which appeared in print in 1620: 23027̊0022, where the ∘ indicates the position of the units. In 1616 Kepler wrote "43 (18" for 43·18, but in 1624 he adopted

Magini's notation. Napier also employed various different notations, and a number of very different systems were in fact in use throughout the seventeenth century.

Even at the beginning of the eighteenth century Father Reyneau writes "809·132275^{VI}", using the Roman numerals to indicate the number of decimal places, and in 1702 Manesson Mallet writes "30 toises* 5′ 6″ 9‴ 4⁗ 3‴‴" for what we should write as 30.56945 toises. Mallet's notation is like that used by Jean Boulenger in 1630.

Stevin not only recommends that decimal numbers should be used for calculations but also suggests that units of measurement should be divided decimally:

THE FIFTH ARTICLE OF ASTRONOMICAL COMPUTATIONS

The ancient astronomers having divided their circles each into 360 *degrees*, they saw that the astronomical computations of them with their parts was too laborious; and therefore they divided also each degree into certain parts, and these again into as many, etc., to the end thereby to work always by whole numbers, choosing the 60th progression because that 60 is a number measurable by many whole measures namely 1, 2, 3, 4, 5, 6, 10, 12, 15, 20, 30; but if experience may be credited (we say with reverence to the venerable antiquity and moved with the common utility), the 60th progression was not the most convenient (at least) amongst those that in nature consist potentially, but the tenth, which is thus. We call the 360 degrees also *commencements*, expressing them so 360 ⓪, and each of them a degree or 1 ⓪ to be divided into 10 equal parts, of which each shall make 1 ①, and again each 1 ① into 10 ②, and so of the rest, as the like hath already been often done.

Now this division being understood, we may describe more easily that we promised in addition, subtraction, multiplication, and division; but because there is no difference between the operation of these and the four former propositions of this book, it would but be loss of time, and therefore they shall serve for examples of this article; yet adding thus much that we will use this manner of partition in all the tables & computations which happen in astronomy, such as we hope to divulge in our vulgar German language, which is the most rich adorned and perfect tongue of all other, & of the most singularity, of which we attend a more abundant demonstration than Peter and John have made thereof in the Bewysconst and Dialectique, lately divulged.

---

* "toise" = fathom. J.V.F.

The new decimal division of the degree was adopted by Briggs in his tables called *Trigonometria Britannica* printed in Gouda in 1633.*

However, Vlacq's *Trigonometria Artificialis*, which was printed in the same city and in the same year, proved more popular than Briggs's work, being better presented. It used sexagesimal divisions of the degree.

When the National Convention set up after the French Revolution worked out the metric system of units French astronomers and mathematicians showed renewed interest in Stevin's idea, though they adopted centesimal rather than decimal divisions, defining the right angle as 100.

Difficulties arose from the fact that the angle of the equilateral triangle, which was adopted as a unit by the Babylonians and plays a very important role in geometry, could not be expressed exactly in the new system.

Now, two-and-a-half centuries after Stevin, astronomers are beginning to show an interest in the system he proposed.

Stevin was not, in fact, merely concerned with trigonometry; he also wanted to change the whole system of measurements.

In his *Practical Arithmetic* of 1611 Adriaan Metius stresses the advantages of using decimal fractions for calculation. In 1630 Jean Boulenger, a professor at the Collège Royal adopted Stevin's technique in his textbook of practical geometry. But he was obliged to give graphs which would allow him to convert from conventional units into decimal ones at the beginning of his calculations and then back again to conventional units at the end.

A decimal system of measurement was not introduced until after the French Revolution.

On 11 Floréal† of the Year III‡ Laplace announced at the

---

* He says the idea was suggested to him by Vieta.

† April/May, the eighth month of the French Republican calendar.

‡ 1794.

École Normale:

Today I am going to depart from the mathematical curriculum to discuss the system of weights and measures which has just been definitively introduced by a decree of the National Convention. When you return to the provinces it will be one of your most important functions to explain this victory for science and for the Revolution to your fellow-citizens, and especially to the teachers in primary schools. This system is so very important that I shall now describe it in considerable detail.

An enormous number of different units of measurement is in use. The units vary not only from nation to nation but also from section to section of the same community. The units are subdivided in illogical ways which make calculations difficult: the divisions are difficult to remember and difficult to compare among themselves. In commerce such units are inconvenient, and they make fraud easier. So one of the greatest services that scientists and the Government can render society is to adopt a system of measurement with a standard subdivision of units, designed to facilitate calculation, and with units derived, in the least arbitrary manner possible, from some fundamental property of Nature. A nation which adopts such a system of measurement not only will be the first to benefit from it but will also see other nations follow its beneficent example. The slow but irresistible advance of Reason eventually triumphs over national rivalry and over all the obstacles to the attainment of any end which is generally recognized as desirable. Such were the motives of the Constituent Assembly in entrusting the Academy of Sciences with this important task. The new system of weights and measures is the result of the work carried out by members of the Academy's Commission, helped and guided by the dedicated efforts of several representatives of the Nation.

. . . . .

The Academy Commission suggested that to make it easier to calculate the gold and silver content of coins, they should be made using a ten per cent alloy and their weights should be decimal multiples of a gramme. Finally, they decided that for the sake of uniformity within the system the day must be divided into ten hours, the hour into one hundred minutes, the minute into one hundred seconds, etc....

Such a division of the day, which will be necessary for astronomers, is less useful in everyday life, where we rarely need to multiply or divide by time. The difficulty of using this system on clocks and watches, and the fact that we trade in such instruments with other nations have caused us to postpone indefinitely the acceptance of this recommendation. It is likely, however, that in the long term a decimal division of the day will replace the present system, which must eventually be abandoned because it is so much at variance with the system used for all the other units.

Such is the new system of weights and measures the members of the Academy suggested to the National Convention, which immediately adopted it. Since the system is based on the length of the terrestrial meridian it applies equally to all peoples. The only relation with France is in the choice of a meridian which passes through this country, but this particular meridian has the advantages of having both its ends in the sea and of being cut by the mean parallel, which would also have sufficed to recommend its use as a universal standard to an international assembly of scientists. We may thus permit ourselves to look forward to the day when this system is adopted everywhere. Since it is incomparably simpler than the old system in its divisions and in its nomenclature children will find it much easier. You will find it harder to explain to teachers, who have long been familiar with the old measures: they will find the new system complicated, because it is natural to think something is complicated when our prejudices and habits make it difficult for us to understand it. However, your intelligence and dedication will overcome these obstacles.

The difficulties foreseen by Laplace did in fact arise, and the metre and other units were not definitively fixed until the 19 Frimaire of the Year VIII (10 December 1799). Article 4 of the law being this date states: "A medal shall be struck to preserve for posterity the date at which the metric system was brought to perfection and to explain the operation which forms its basis. The inscription on the obverse of the medal is to be 'A tous les temps, à tous les peuples' ['To all times and to all peoples'] and on the reverse 'République française, an VIII' ['Republic of France, Year VIII']."

However, the definitive metric system was not introduced until 2 November 1801.

The public proved to be recalcitrant and clung to its old ways. As early as the 13 Brumaire of Year IX (4 November 1800), a year before the metric system was definitively adopted, it was agreed that the new units might be called by the old names. For example, the hectogramme might be called an *once* [ounce],* although its value was more than three *onces*, the name *livre* [pound] might be given to the kilogramme, though the new unit was more than twice as large as the old one, and the myriamètre might be called a

---

* The name in brackets is the equivalent English unit. J.V.F.

lieue [league], though its length was two and a quarter *lieues*, etc. . . .

Retail shopkeepers traded on the ignorance of their customers and continually cheated them. The Imperial Government took action to stop such practices; on the 12 February 1812 it promulgated a decree, brought into force on 28 May, to authorize the use of a mixed system, which lasted until 1840.

For commercial purposes it was permitted to use a measuring rod two metres long divided into six *pieds* [feet], the pied was divided into twelve *pouces* [inches] each divided into twelve *lignes* [lines]. Each measuring rod must have two scales: one side showing the old divisions and the other side showing the decimal divisions. Similar regulations applied for measures of volume and of weight.

However, the provisions of the decree of 1812 only applied to retail trade: the official units were to be used for public works and for wholesale trade, and the official system was to be the only one taught in the state schools.

In 1837, on a motion proposed by the Minister of Trade, the French Parliament rescinded the decree of 1812 and ordered that from 1 January 1840 all weights and measures other than those established by the laws of 18 Germinal, Year III,* and 19 Frimaire, Year VIII,† were to be illegal, and the use of them was to carry the penalties provided by article 479 of the Penal Code.

Thus it was only in the mid-nineteenth century, hardly more than a century ago, that Stevin's ideas were at last accepted in France. The Anglo-Saxon countries have not yet completely accepted them.

---

* March/April, 1794.

† November/December, 1799.

2. Algebraic Notations and First Degree Problems

*On apportait des cartes, non pour jouer,
mais pour y apprendre mille petites gentil-
lesses et inventions nouvelles, lesquelles
toutes issaient d'arithmétique. En ce moyen
entra en affection d'icelle science numérale,
et, tous les jours après dîner et souper, y
passait temps aussi plaisantement qu'il
soulait ès dés ou ès cartes. A tant sut d'icelle
et théorique et pratique, si bien que Tunstal,
Anglais qui en avait amplement écrit,
confessa que vraiment, en comparaison de
lui, il n'y entendait que le haut allemand.**
Rabelais

2.1. Notations

Today mathematicians make use of a wide variety of conventional notations without which they would find it impossible to write, or even think, about any mathematical problem.

Historically, however, this use of symbols is a relatively recent phenomenon.

The Egyptians used a certain number of specifically mathematical ideograms: for instance, *addition* was

---

* Cards were brought in, not to play with, but so that he might learn a thousand little tricks and new inventions, all based on arithmetic. In this way he came to have this science of numbers, and every day, after dinner and supper, whiled away the time with it as pleasantly as formerly he had done with dice and cards, and so he came to know the theory and practice of arithmetic so well that Tunstad, the Englishman who had written so copiously on the subject, confessed that really, in comparison with Gargantua, all that he knew of it was so much nonsense. Trans.: J. M. Cohen.

indicated by a pair of legs walking in one direction $\boxed{\text{ЛЛ}}$, and *subtraction* by a pair of legs walking in the opposite direction $\boxed{\text{ЛЛ}}$, etc. This is probably explained by the fact that Egyptian writing, like that of the Babylonians, contained a large number of ideograms.

The Greek mathematicians of the third century BC did not use any symbols at all: every line of the argument was written out in words. Diophantus, however, introduced a few abbreviations. *Subtraction* is indicated by $\boxed{\lambda}$, the unknown by a sign which Paul Tannery gives as $\boxed{\varsigma}$, the square of an unknown is indicated by $\boxed{\Delta^Y}$ ($\delta\bar{\upsilon}\nu\alpha\mu\iota\varsigma$, power), its cube by $\boxed{\kappa^Y}$ ($\kappa\upsilon\beta o\varsigma$, cube), its fourth power by $\boxed{\Delta^Y_\Delta}$ ($\delta\bar{\upsilon}\nu\alpha\mu o\delta\bar{\upsilon}\nu\alpha\mu\varsigma$, power power, square square), its fifth power by $\boxed{\Delta\kappa^Y}$ ($\delta\bar{\upsilon}\nu\alpha\mu o\kappa\upsilon\beta o\varsigma$, square cube) and finally its sixth power by $\boxed{\kappa^Y\kappa}$ ($\kappa\upsilon\beta o\kappa\upsilon\beta o\varsigma$, cube cube). Diophantus worked only with rational numbers, integral or fractional. He used symbols to denote the reciprocals of the above unknowns:

$\boxed{\varsigma^x}$ $\;\alpha\rho\iota\theta\mu o\sigma\tau\acute{o}\nu$, reciprocal of the unknown,

$\boxed{\Delta^{Yx}}$ $\;\delta\bar{\upsilon}\nu\alpha\mu o\sigma\tau\acute{o}\nu$, reciprocal of the square, etc. . . . Addition is indicated by simple juxtaposition. Unity is written $\overset{\circ}{\text{M}}$.

The polynomial $x^3 - 5x^2 + 8x - 1$ is written:

$$\boxed{\text{K}\overset{Y}{\alpha}\varsigma\eta\lambda\Delta^Y\epsilon\overset{\circ}{\text{M}}\alpha}$$

(cube 1 unknown 8 minus squares 5 units 1).

The quotient is written $\acute{\epsilon}\nu\ \mu o\rho\iota\omega$: then

$$\boxed{\Delta^Y\tfrac{\xi}{}\overset{\circ}{\text{M}},\beta\phi\kappa\acute{\epsilon}\nu\mu o\varsigma\acute{\iota}\,\omega\overset{Y}{\Delta}\Delta\alpha\overset{\circ}{\text{M}}\lambda\lambda\Delta^Y\varsigma}$$

is to be read as

$$\frac{60x^2 + 2520}{x^4 + 900 - 6x^2}.$$

Diophantus never uses more than one unknown.

Hindu mathematicians frequently employed symbols, and sometimes used several unknowns, but later on the Arabs nevertheless wrote out their equations in words, as also did Italians such as Leonardo of Pisa (in the thirteenth century), Luca Pacioli (in the fifteenth) and Cardan and Tartaglia (in the sixteenth), though the last used some slight abbreviations.

Symbols do however appear in French and German work of the fifteenth and sixteenth centuries.

These symbols vary a great deal from one author to another and two different kinds of terminology are used.

$$1, x, x^2, x^3, x^4, x^5, x^6, x^7$$

is read as: unity, side (or root), square, cube, square of square, first relative, square of cube, second relative, etc. . . .*
and is written by Henrion (who provides us with a late example of this usage) as

$$N, R, q, c, qq, \beta, qc, b\beta, \text{etc.} \dots$$

In the other kind of terminology, derived from Diophantus, whose work was known in Xylander's Latin translation of 1575,

$$1, x, x^2, x^3, x^4, x^5, x^6,$$

is read: unity, side, square, cube, square square (or double-square), square cube, cube cube, and is written $1, N, Q, C, QQ, QC, CC$, etc. . . .

---

* This nomenclature, which is used by the Arabs and by Luca Pacioli, dates back to the Greeks, more specifically to Anatolius of Alexandria, Bishop of Laodicea (second half of the third century AD). In Vol. 1, p. 269, we show the notation Tartaglia used with his arithmetic triangle. The alternative system which we shall describe below also derived from Greek work.

Nicolas Chuquet, however, uses an elegant and very suggestive system.

When we should write: Chuquet writes:

$$12x$$ $$12^1$$
$$13x^2$$ $$13^2$$
$$x^3$$ $$1^3$$
$$12x^{-1} \text{ or } 12/x$$ $$12^{1m}$$
$$12$$ $$12^0$$ to indicate that x does not appear.

For roots Chuquet uses the symbol R:

It is convenient to take any number standing alone to signify the first root, so that for the first root of . 12 . which we may write by putting 1 above R like this: R^1, R^1. 12 . is . 12 . And R^1. 9 . is . 9 . and so on for all other numbers. The second root is that number which when written in two positions one above the other which are then multiplied by one another gives a product equal to the number whose second root it is. Such as . 4 . and . 4 . which multiplied one by the other give . 16 . So the second root of . 16 . is . 4 .... We can also write it R^2. 16 .... The sixth root should thus be written R^6 ...

... There are other types of roots beside the simple ones we have mentioned above, and we may call these compound roots. For example . 14 . plus R^2. 180 . whose second root is . 3 . p . R^2. 5 ... the second root of 14 . p . R^2. 180 . can thus be written $R^2$14 . p . R^2. 180 .

This last passage means that

$$\sqrt{14 + \sqrt{180}} = 3 + \sqrt{5}.$$

What we should write as $\sqrt{4x^2 + 4x} + 2x + 1 = 100$ Chuquet writes

$$R^2 \;\underline{4^2 \text{ p } 4^1} \text{ p } 2^1 \text{ p } 1 \text{ equal to } 100.$$

Chuquet's use of exponents is a considerable advance upon previous systems and is in fact as convenient as the modern system. For example, modern notation makes it obvious that

$$4x^2 \times 6x^3 = 24x^5.$$

This is also obvious from Chuquet's notation, but an author who writes "4 q multiplied by 6 c" needs to know the result of multiplying a square by a cube. Diophantus's terminology gives 24 qc which is not excessively clumsy, but the Arab system requires the user to know that a square multiplied by a cube gives a first relative, and the product is then written 24 β.

Tartaglia gives a detailed account of what is involved: we find from a table that q is the second power, c the third. We apply a theorem which goes back to Archimedes: the product of the second power by the third power is the fifth power $(2 + 3 = 5)$. From the table we find that the fifth power is β. Chuquet's notation makes such translations unnecessary, and it is not surprising that, having been used by Bombelli, Stevin, Albert Girard, Kepler and Descartes, among others, it was eventually adopted universally.

French and Italian mathematicians continued for some time to use the abbreviations p and m to indicate addition and subtraction. The modern $+$ and $-$ signs, which first appeared in a work on arithmetic by Johann Widmann, gained currency after Michael Stifel used them in his famous *Arithmetica Integra* of 1544.

The modern "equals" sign, $=$, was introduced by the Englishman Robert Recorde in 1557, but in the seventeenth century it was still possible for Descartes to use a different sign: ∞.

The "greater than" sign ($>$) is due to Harriot, the multiplication sign ($\times$) to Oughtred, the root sign ($\sqrt{}$) to Rudolff, etc.... Such developments were not, however, of vital importance. Until this time algebraic work had been numerical and had usually involved only one *unknown*.

Algebraists who wanted to use symbolic notation in work with two or more unknowns found it natural to indicate the different unknowns by different symbols.

Stifel represented the first unknown in the ordinary way and then used A, B etc., to indicate the further unknown quantities. This notation was simple and convenient.

In his work on first degree problems, published in 1559, the French mathematician Borrel (or Buteo) uses notation very like that of Stifel. (See Plate IV, p. 59.)

By the middle of the sixteenth century elementary algebra had thus reached something like its modern state of development.

The important step was to be taken by Vieta. Algebra had always been used in dealing with geometrical problems, but it presented several disadvantages: the problem had to be stated numerically, and preferably in terms of simple numbers so that the complexity of the calculations should not obscure the outline of the reasoning. However careful one was, it was always difficult, at the end of the calculations, to trace the line of the argument. Vieta made the procedure much clearer by deciding:

1. To use letters of the alphabet $A, B, C \dots$ to represent all the quantities used in the calculations, whether their values were known or not.

(In this he was merely following the tradition of the Greek geometers.)

2. To keep the consonants B, C, D, F, etc. . . . . for known quantities and use the vowels A, E, O, etc. . . . for unknowns.

3. To indicate geometric or arithmetic addition by $A + B$ (the modern notation), subtraction by $A - B$ if A is greater than B, or by $A = B$ if we are concerned with the absolute value of a difference and it is not known which of the quantities is the greater. A product is written $\underline{A\ in\ B}$ (A into B), a quotient A/B.

4. The letters all represent geometrical quantities whose type is always indicated, so that Vieta can ensure that all his equations are homogeneous. For example, the equation we should write as $b^2 + d = c$ is written by Vieta as "Proponatur B in A quadratum, plus D plano in A, aequari Z solido." "It is proposed to make B into A squared plus D plane into A equal to the solid Z."

The expression

$$\frac{df + d^2 - b^2}{2(f + d)b}$$

is written

$$\frac{D \text{ in } F + D \text{ quadr} - B \text{ quadr}}{F + D, \text{ in } B \text{ bis}}$$

and in this case the dimensions are not indicated. The context makes it clear that we are concerned with lengths.

Although geometry only admits three dimensions Vieta does not hesitate to use quantities of dimension four, five or even more.

Vieta's system proved to be a powerful tool, though his notation was simplified by the mathematicians of the first part of the seventeenth century.

Descartes uses something very like modern notation: a letter no longer represents a quantity but instead its ratio to a quantity of the same kind chosen as a unit. This enables Descartes to avoid geometrical quantities of more than three dimensions. Upper-case letters are replaced by lowercase ones; and Diophantus's system of powers is replaced by that of Chuquet.

Descartes uses the modern form $\sqrt{a^2 + b^2}$ for square roots but writes the cube root as $\sqrt{C. a^3 - b^3 + ab^2}$.

For unknowns Descartes uses not the vowels but the last letters of the alphabet. He prefers, however, to start his unknowns at z rather than at x: designating the first unknown by the last letter of the alphabet, the second by the second to last letter and so on.

He avoids using negative quantities, which he calls "false", though he does not forbid their use.

In the main, however, the publication of Descartes's *Geometry* of 1637 marks the point when the notation of elementary algebra assumed its present form.

2.1. The first degree

We shall now consider some problems which would today be classified as of the first degree. Certain of them are very elementary, and many date back to remote antiquity.

Let us start with an Egyptian problem from the Rhind papyrus:

100 loaves among 5 persons. The last two receive $\frac{1}{7}$ of what the first three receive. What is the difference?

The solution (which we quote below) shows that the loaves are divided according to an arithmetic progression. If the person who receives least receives x loaves, the others will receive $x + y$, $x + 2y$, $x + 3y$, $x + 4y$. The total is $5x + 10y = 100$, therefore $x + 2y = 20$ (1).

Also, the first three altogether receive $3x + 9y$ while the other two receive $2x + y$, so the second condition gives

$$3x + 9y = 7(2x + y)$$

i.e.

$$11x = 2y \qquad (2).$$

Substituting in (1) we obtain $12x = 20$

$$\therefore \quad x = \tfrac{20}{12} = \tfrac{5}{3} = 1 + \tfrac{2}{3},$$

and

$$y = \tfrac{5}{3} \times \tfrac{11}{2} = \tfrac{55}{6} = 9 + \tfrac{1}{6}.$$

The solution given in the Rhind papyrus is as follows:

Do as it occurs: difference $5\frac{1}{2}$

$$23, 17\frac{1}{2}, 12, 6\frac{1}{2}, 1.$$

Increase the numbers by 1 times $\frac{2}{3}$; this gives for

23	$38\frac{1}{3}$
$17\frac{1}{2}$	$29\frac{1}{6}$
12	20
$6\frac{1}{2}$	$10\frac{2}{3}\frac{1}{6}$
1	$1\frac{2}{3}$
Sum 60	Sum 100.

It is very difficult to reconstruct the reasoning behind this table of figures.

A similar early Babylonian problem [1] reads as follows:

10 brothers. 1 mina and two thirds of a mina of silver. One brother is raised above the other. How much he is raised is unknown. The eighth receives 6 shekels. How much is one brother raised above the other?

We are again dealing with numbers in an arithmetic progression with an unknown common difference: "one brother is raised above the other". We are required to find the common difference of the series.

The Babylonian scribe gives the following solution:

You, by calculating, find the reciprocal of 10, the number of men, which gives 6′. You form the product of 6′ with 1 mina and two thirds of silver which gives 10′. Double 10′, which gives 20′. Double 6′, the share received by the eighth brother, giving 12′. Subtract 12′ from 20′ leaving 8′. Remember the number 8′. Add 1 above to 1 below, which gives 2. Double 2, which gives you 4. You add 1 to 4, which gives you 5. Subtract 5 from 10, the number of men, which gives you 5. Find the reciprocal of 5, which is 12′. Multiply 12′ by 8′ which gives you 1′ 36″: this is the amount by which one brother is raised above the other.

First of all, we must remember that the calculations are being carried out in a sexagesimal system of arithmetic.

The units of weight are also sexagesimal. In decreasing order of size they are the talent, the mina, the shekel and the grain.

We shall express the problem in the form of an equation. Let the common difference of the progression be x. Then since the eighth brother receives 6 shekels the shares received by the others will be as follows: the ninth will receive $6 - x$, the tenth $6 - 2x$, the seventh $6 + x$, and the preceding ones $6 + 2x$, $6 + 3x$, $6 + 4x$, $6 + 5x$, $6 + 6x$ and $6 + 7x$.

The sum is $60 + 25x$. This must come to 1 mina and $\frac{2}{3}$, or 100 shekels.

So $60 + 25x = 100$

$\therefore$ $25x = 40$

and

$$x = \tfrac{40}{25} = \tfrac{8}{5} = 1\tfrac{3}{5} \text{ or } 1°36' \text{ in shekels,}$$

making $1'36''$ in minas.

This result is the same as that obtained on the tablet.

Another Babylonian problem [2] reads:

Per *bur*, I have obtained 4 *kur* of grain. From another field I have obtained 3 *kur* of grain per *bur*. One lot of grain is 8'20 greater than the other. I have added up my fields: they come to 30'. What size are they?

The Sumerian unit of area was the *sar*, and the *bur* is a larger unit equal to 30' i.e. 30 second-order units, or 1800 *sar*. The *kur* is a unit of volume equal to 5' i.e. 300 of the basic unit the *qa*. [3]*

There are two unknowns in this problem. The proprietor of two fields leases them, the first at 4 *kur* of grain per *bur* (as we might now lease a field at 4 hundredweight of corn per acre), and the second at only 3 *kur* per *bur*. The first field yields 8'20 *qa* more than the second. The total area of the fields is 30'.

Since the mean area is 15' the first field will have an area $15' + x$ and the second an area $15' - x$. We now have only one unknown.

---

* Here, units are indicated by °, negative powers of 60 by ', ", etc., and positive powers of 60 by ', ", etc.

Since 1 *bur* = 30` *sar*, one *sar* of the first field produces

$$\frac{5` \times 4}{30`} = 40' \ qa.$$

One *sar* of the second field produces

$$\frac{5` \times 3}{30`} = 30' \ qa.$$

Therefore

$$(15` + x) \times 40' - (15` - x) \times 30' = 8`20$$

that is

$$10` + 40'x - 7`30 + 30'x = 8`20$$

$$2`30° + 1°10'x = 8`20$$

$$1°10'x = 5`50$$

$$x = \frac{5`50}{1°10'} = 5`.$$

The first field therefore has area 15` + 5` = 20` and the second 15` − 5` = 10`.

The Babylonian scribe employs exactly this method:

Write down 30`, the *bur*. Write 20`, the grain produced. Write 30`, the second *bur*. Write 15`, the grain produced. Write 8`20, the amount by which one amount of grain exceeds the other. Finally write 30`, the sum of the areas of the fields.

Then divide 30`, the sum of the areas of the field, by two: 15`. Write 15` and 15` twice. Find the reciprocal of 30`, the *bur*: 2″. Multiply 2″ by 20`, the grain it produced: 40', the false grain. Multiply this by 15`, which you have written down twice: 10`. Remember this number. Find the reciprocal of 30`, the second *bur*: 2″. Multiply 2″ by 15`, the grain produced: 30', the false grain. Multiply by 15`, which you have written down twice: 7`30. By how much does 10`, the number you remembered, exceed 7`30? By 2`30. Subtract 2`30, the difference, from 8`20, the amount by which one amount of grain exceeds the other: this leaves 5`50. Remember the remainder of 5`50. Add the coefficient 40' and the coefficient 30': 1°10'. I do not know its reciprocal. By what must I multiply 1°10' to get 5`50, the remainder I had to remember? Write 5'. Multiply 5' by 1°10', this will give you 5'50. Using the two 15''s you wrote, subtract from one and add to the other 5', the number you have written down: the first gives 20', the second 10'. The area of the first field is 20', the area of the second is 10'.

The solution is immediately followed by a check:

If the area of the first field is 20' and the area of the second field 10', how much grain do they produce? Find the reciprocal of 30', the *bur*: 2''. Multiply 2'' by 20', the grain it produced: 40'. Multiply by 20' the area of the first field: 13'20, the grain from 20', the area of the first field. Find the reciprocal of 30', the second *bur*: 2''. Multiply 2'' by 15', the grain it produced: 30'. Multiply 30' by 10', the area of the second field: 5', the grain from 10', the area of the second field. By how much does 13'20, the grain from the first field, exceed 5', the grain from the second field? By 8'20.

It will perhaps be instructive to examine an early form of another problem, which is also of the first degree but rather more sophisticated since it involves three unknowns.

We shall begin by considering the problem as it appears in the work of Diophantus, Book I, Problem 24.*

To find three numbers such that if we add to each of them a fraction of the sum of the other two the resultant totals are equal.

Suppose the problem to be as follows: if we add to the first number one third of the sum of the other two numbers, to the second number one quarter of the sum of the other two numbers and to the third number one fifth of the sum of the other two numbers, these three totals are to be equal.

Let the first number be $1N$, and, since we are to add to it one third of the sum of the remaining numbers, let this sum (for convenience) be a number which can be exactly divided into thirds, namely 3. Now the sum of the three numbers is $1N + 3$ and the total obtained by adding the first number to one third of the sum of the other two is $1N + 1$.

We also require that the second number plus one quarter of the sum of the other two numbers shall be $1N + 1$. Let us consider four times this total. Thus four times the second number plus the two other numbers gives three times the second number plus all three of the numbers. So three times the second number plus all three numbers gives $4N + 4$. Now if we subtract the sum of the three numbers from each side the remainder of $3N + 1$ is equal to three times the second number and the second number will be $1N + \frac{1}{3}$.

---

* We have made some modifications to the solution, in particular by adopting the notation used by Bachet.

We further require that the third number plus a fifth of the sum of the other two shall be $1N + 1$. Using the method of the preceding paragraph we consider five times this total and deduce that the third number is $1N + \frac{1}{2}$.

Finally, we require that the sum of the three numbers shall be $1N + 3$, so $1N$ is $\frac{13}{2}$. Now, eliminating the fraction, we find that the first number is 13, the second 17, and the third 19. These numbers satisfy the required conditions.

The problem has an infinite number of solutions, since any multiple of a solution is also a solution.

Diophantus takes advantage of this fact to stipulate that the sum of the last two numbers shall be 3 units. This does not affect the generality of the solution, since the three numbers obtained may be multiplied by any factor we please once they have been calculated.

Diophantus writes out his equations in words, but we shall translate them into symbolic form, denoting the three numbers by x, y and z.

Diophantus takes x as his principal unknown, writing it as $1N$.

He then takes $y + z = 3$. The sum of the three numbers will therefore be $x + 3$.

So

$$x + \frac{y + z}{3} = x + 1.$$

The second condition requires that

$$y + \frac{x + z}{4} = x + 1$$

that is

$$4y + x + z = 4x + 4$$

$$3y + (x + y + z) = 4x + 4.$$

Since the sum of the numbers is $x + 3$ we have:

$$3y = 3x + 1, \quad \text{and} \quad y = x + \tfrac{1}{3}.$$

The third condition gives:

$$z + \frac{x + y}{5} = x + 1$$

$$5z + x + y = 5x + 5$$

i.e.,

$$4z + (x + y + z) = 5x + 5$$

$$\therefore \quad 4z = 4x + 2, \quad z = x + \tfrac{1}{2}.$$

Substituting these values for y and z into the equation $y + z = 3$ we obtain:

$$x + \tfrac{1}{3} + x + \tfrac{1}{2} = 3$$

$$2x = 3 - \tfrac{5}{6} = \tfrac{13}{6},$$

$$x = \tfrac{13}{12},$$

and thus

$$y = \tfrac{13}{12} + \tfrac{1}{3} = \tfrac{17}{12}; \quad \text{and} \quad z = \tfrac{13}{12} + \tfrac{1}{2} = \tfrac{19}{12}.$$

The three numbers required are therefore in the ratio 13:17:19.

The same problem, involving slightly different numbers and one more condition, is to be found in two late fifteenth-century works: the *Summa* of Luca Pacioli and the treatise on arithmetic of Pietro Borgi of Venice, as well as in Tartaglia's *General Trattato*, which was written in the sixteenth century. All three authors solve the problem by the method of double false hypothesis.*

Tartaglia stated the problem as follows (following the translation made by Guillaume Gosselin in 1578):

Three merry companions, with money in their purses, were conversing with one another. The first said to the other two, "If you give me half your ducats I shall have, together with what I have already, 20 ducats". The

_____
* See below, p. 55.

second said to the other two, "If you give me a third of your ducats I shall have, together with what I have already, 20 ducats". "But", said the third to the other two, "give me a quarter of what you have and with what I have got already I shall have 20 ducats as well". How many ducats did each of them have?

If we follow Diophantus's method and use only the fact that the three totals are equal, we obtain the result that the three shares are in the ratio $5:11:13$. If we now write them as $5x$, $11x$ and $13x$ the first condition gives us

$$5x + \frac{11x + 13x}{2} = 20 \, ,$$

that is

$$5x + 12x = 20$$
$$17x = 40$$
$$x = \tfrac{40}{17} \, ,$$

from which we obtain the results given by Pacioli, Borgi and Tartaglia, namely: $5\tfrac{15}{17}$, $12\tfrac{16}{17}$, $15\tfrac{5}{17}$.

However, these authors approach the problem in a manner very different from that of Diophantus: they use the Arab method of false hypothesis.

Borgi, in 1501, starts by assuming that the first of the friends gets 12 ducats. He then sets about satisfying the first two conditions.

The first condition is that the two other friends must together have twice $(20 - 12)$, i.e. 16. Borgi supposes the second of the friends has 7, and the third therefore has 9.

This satisfies the first condition. As for the second one: $7 + \frac{12 + 9}{3} = 14$, whereas we require 20. Our number is too small by $20 - 14 = 6$.

Let us therefore alter our assumptions. The first friend is still to have 12, but we shall now suppose the second to have 10, and the third therefore has 6. The second condition gives: $10 + \dfrac{12 + 6}{3} = 16$, whereas we require 20. Our answer is too small by $20 - 16 = 4$.

Assuming the first friend has 12, the second's share is now worked out by the classical argument of double position (that is linear extrapolation)

For 7 the error is   6
For 10 the error is   4
For x the error is   zero.

Therefore
$$\frac{x - 10}{10 - 7} = \frac{4}{6 - 4},$$

$$x = 16.$$

The traditional procedure does not involve reasoning. It merely supplies a formula to be used at this point: cross-multiply, $7 \times 4 = 28$, $10 \times 6 = 60$, subtract one result from the other, $60 - 28 = 32$, and divide by the difference between the errors, $6 - 4 = 2$, which gives $x = \frac{32}{2} = 16$.

So if we assume that the first number is 12, the first two conditions can be satisfied by taking the second number to be 16. Burgi does not mind that the third number is then zero.

However, is the third condition satisfied?
$0 + \dfrac{16 + 12}{4} = 7$, which is too small by $20 - 7 = 13$.

We go back to the beginning and this time assume that the first man has 9 instead of 12. Half the sum of the shares of the other two is $20 - 9 = 11$, so these two shares add up to 22.

Suppose that the second man has 7. and the third therefore has 15. What does the second condition give us? $7 + \dfrac{9 + 15}{3} = 15$, whereas we require 20. The number is too small by 5. Let us suppose the second man has 10, and the third therefore has 12: $10 + \dfrac{9 + 12}{3} = 17$, too small by 3. The second man's share

		error
7	too small by	5
10	too small by	$\frac{3}{2}$

from which we obtain (by taking the difference of the cross products)

$$\frac{50 - 21}{2} = \frac{29}{2} = 14\tfrac{1}{2}.$$

The second man has $14\tfrac{1}{2}$, and the third one $7\tfrac{1}{2}$, supposing the first man to have 9.

Third condition:

$$7\tfrac{1}{2} + \frac{9 + 14\tfrac{1}{2}}{4} = 13\tfrac{3}{8}, \qquad \text{too small by } 6\tfrac{5}{8}.$$

Having assumed two different values for the first man's share Borgi can now work out what it should be by applying the technique of false hypothesis:

First man's share	error	
12	13	$13 \times 9 = 117$
9	$6\tfrac{5}{8}$	$12 \times 6\tfrac{5}{8} = 79\tfrac{1}{2}$
	$6\tfrac{3}{8}$	$37\tfrac{1}{2}$

The first man's share is $37\frac{1}{2}/6\frac{3}{8} = 5\frac{15}{17}$. The same procedure gives us the second's share:

Second man's share	error	
16	13	$14\frac{1}{2} \times 13 = 188\frac{1}{2}$
$14\frac{1}{2}$	$6\frac{5}{8}$	$6\frac{5}{8} \times 16 = 106$
	$6\frac{3}{8}$	$82\frac{1}{2}$

The second man's share is $82\frac{1}{2}/6\frac{3}{8} = 12\frac{16}{17}$. The same procedure is used again to work out the third man's share, which comes to $15\frac{5}{17}$.

In his translation of Tartaglia Gosselin uses a mixed method. He adds:

> So we have explained this difficult problem in a simple manner, but we could nevertheless deal with it more easily and more briefly by using algebra, either by the rule of simple quantities or by the rule of surds, as we have explained in our algebra, where we discussed these rules at length; therefore we shall postpone such fuller and more subtle explanations until we come to treat of numbers more arcane and divine than these, from whose properties these two rules of position are derived, as we have shown in full in our algebra.

Gosselin gives an elegant and rigorous justification of the two methods of Helcatayn (single and double false hypothesis). Both Vieta, in his *Zetetics*, and Prestet in his *New Elements* (*Nouveaux Eléments*), returned to Diophantus's problem and solved it algebraically. It also reappears in the work of Saunderson in the eighteenth century, and even in Euler's *Algebra*. A noteworthy feature of the problem as it appeared in Borrel's *Logistics* (1559) (Plate IV) is the algebraic notation which, though still primitive in some respects, is very elegant. Our translation uses rather more sophisticated notation:

> Given a certain sum, find three numbers such that the first plus half the sum of the others, the second plus one third of the sum of the others and the third plus one quarter of the sum of the others each add up to the given sum.
>
> Let the given sum be 17. Let us write the first of the required numbers as a, the second as b and the third as c.

192 *LIBER*

C, *& secundus* B. *Et ita cùm in æquatione postre*
ma,ex duobus numeris antecedentibus, alter fue-
rit monas, residuum, & proueniens erunt duo ex
quæsitis numeris.Quod tamen aliquando fallit,sed
rarißimè.Multis præterea modis super factus æqua
tionibus ratio procedet,quorum erit vtilior studio-
sis inuestigatio propria,quàm aliena traditio.

Data summa qualibet, tres numeros in-
uenire,quorum primus cum semisse, secun
dus cum triente, tertius cum quadrante re-
liquorum eam summã singuli constituant.

E *sto data summa* 17. *Pone primum ex nume*
ris quæsitis esse 1 A,*secundum* 1 B, *tertium*
1 C. *Erit igitur* 1 A,$\frac{1}{2}$ B,$\frac{1}{2}$ C [17. *Item* 1 B,
$\frac{1}{3}$ A,$\frac{1}{3}$ C [17. *Et etiam* 1 C $\frac{1}{4}$ A, $\frac{1}{4}$ B
[17.*Et per æquationem*
secundam,habebis,sicut   2 A. 1 B. 1 C [34
hic ordine collocaui,tres 1 A. 3 B. 1 C [51
æquationes.quas ita tra- 1 A. 1 B. 4 C [68
ctabis. Multiplica ter-
tiam in 2, *fit* 2 A,2 B,8 C [136. *Subducito pri-*
mam,remanet 1 B,7 C [102] *partire in* 7 ,*proue-*
nit 3 *cum residuo* 11,*qui sunt duo numeri , ter-*
tius C & secundus B.Vt habeas primum,ab æqua-
tionis tertiæ numero 68, *detrahe* 4 C,1 B, *id est,*
63,

B.N.

PLATE IV. The elegant algebraic notation of Borrel's *Logistics* (1559).
'Given a certain sum ... (Data summa qualibet ...).

Then we have

$$a + \tfrac{1}{2}b + \tfrac{1}{2}c = 17,$$

$$b + \tfrac{1}{3}a + \tfrac{1}{3}c = 17,$$

and

$$c + \tfrac{1}{4}a + \tfrac{1}{4}b = 17.$$

And writing out these equations again, in order, we obtain

$$2a + b + c = 34$$

$$a + 3b + c = 51$$

$$a + b + 4c = 68.$$

We use them as follows:

The third multiplied by 2 becomes $2a + 2b + 8c = 136$.

Subtracting the first from this we are left with $b + 7c = 102$.

Dividing by 7 we obtain 13, with remainder 11. We thus have two of the numbers: the third, c, and the second, b. To obtain the first we must subtract $4c + 4$, that is 63, from the 68 of the third equation. The result is 5, and this is the first number, a. The three numbers we were required to find are therefore 5, 11 and 13.

If however we want an equation containing only one unknown we must multiply the second equation by 2, which gives

$$2a + 6b + 2c = 102.$$

Subtracting the first equation leaves:

$$5b + c = 68.$$

Multiplying $b + 7c = 102$ by 5 we obtain $5b + 35c = 510$.

Subtracting $5b + c = 68$ we are left with $34c = 442$.

Dividing by 34 gives us 13, which is the third number, c. The other numbers are then found as before.

If, after this, we want to change the given sum, for instance to 85, we proceed as follows: if 17 gives 85, what do we obtain from 5, 11 and 13? Carrying out the calculation we find the numbers 25, 55 and 65, which have properties like those of the previous three numbers.

We shall end our brief survey of first degree problems by quoting statements of some of the most famous of them.

The *Greek Anthology* was assembled some time in the fifth century AD by Metrodorus. Bachet published it in his edition of Diophantus, with a translation into Latin verse. Our quotations are translated from a French adaptation by Montucla.

1.—"Tell me, learned Pythagoras, how many pupils come to your School and listen to your teachings?" "Well," replied the Philosopher, "half of them study mathematics, a quarter study music, a seventh of them keep silence, and there are also three women."

2.—The three Graces, each carrying an equal number of fruits, meet the nine Muses, and each gives them the same number of fruits. Each Grace and each Muse then has an equal number of fruits. How many fruits did each of the Graces originally have?

3.—Water runs into a tank from four pipes. The water coming from one pipe would fill the tank in a day, that from the second would fill it in two days, that from the third in three days, and that from the fourth in four days. How long will it take to fill the tank if water flows in from all four pipes together?

4.—A donkey and a mule were travelling together, and the donkey was complaining: "What are you complaining about?" asked the mule, "If you gave me one of your baskets I should be carrying twice as many as you, and if I gave you one of mine we should each be carrying the same number." How many baskets was each animal carrying?

5.—We have to make a crown, which is to weigh sixty marks,* using gold, copper, iron and tin. The gold and the copper make up $\frac{2}{3}$, the tin and the gold make up $\frac{3}{4}$ and the gold and the iron make up $\frac{3}{5}$. How much is there of each metal?

The next problem concerns the division of a number into parts in some given proportion. It is taken from the *Arithmetic* of Juan Martinez Siliceus, printed in 1526 (in Latin):

Three merchants are in partnership. The first contributed 80 escudos in 6 months, the second 70 escudos in 5 months and the third 60 escudos in 4 months. The total profit was 100 escudos. How should this be divided up fairly, taking into account the differences between the contributions and between the times.

From Tartaglia, following Gosselin:

If 9 workmen drink 12 jugs of wine in 8 days, how much would 24 workmen drink in 30 days?

If 12 oxen eat 3 hundredweight of hay in 15 days, how many oxen would eat 5 hundredweight of hay in 10 days?

---

* One mark = eight ounces.

If 10 labourers can dig up 12 acres of land in 16 days and 12 other labourers can dig up 9 acres in 15 days how many days will it take these 22 labourers to dig up 100 acres of land?

Four men go on pilgrimage together: a gentleman, a workman, a barber and a monk. Their total expenditure is to 10 pounds [livres]. The barber says that he wants to contribute four times as much as the monk, plus four shillings [sols]* more, the workman says that he wants to contribute three times as much as the barber, plus 16 shillings more, and the gentleman says that he wants to contribute twice as much as the workman, plus 10 shillings more. How much is each of them to contribute?

All these problems are simple and they are to be found in many different forms at different dates and in the works of different authors. For example, the Venetian authors Borgi and Tartaglia give the following variant of the problems involving taps. The form is appropriate for a seafaring people, but the use of proportionality is not really justified:

A ship is equipped with three sails. With the largest it can make a certain voyage in 2 days. With the medium sized sail the voyage takes 3 days, and with the smallest sail 4 days. How many days would it take if all the sails were used at once?

In his *Arithmetica Universalis* Newton states a famous problem [4]:

PROBLEM 11. *If cattle* a *should eat up a meadow* b *in time* c, *and cattle* d *an equally fine meadow* e *in time* f, *and if the grass grows at a uniform rate, how many cattle will eat up a similar meadow* g *in time* h?

If a cattle in time c eat up the meadow b, then in proportion (e/b) a cattle in the same time c, or (ec/bf) a cattle in the time f, or (ec/bh) a cattle in the time h, will eat up the meadow e—if, that is, the grass were not to grow after time c. But since, because of the growth of the grass, d cattle in time f eat up only the meadow e, that growth of grass in the meadow e during time $f-c$ will consequently be sufficient enough to pasture $d - (ec/bf)$ a cattle during time f; that is, adequate to pasture $(d - eca/bf)(f/h)$ cattle during time h. And in time $h - c$, proportionately, the growth will be sufficient enough to pasture

$$(df/h - eca/bh)(h - c)/(f - c),$$

---

* The pound (*livre*) was worth 20 shillings (*sols*).

that is,

$$(bdfh - ecah - bdcf + aec^2)/bh(f - c)$$

cattle. Adjoin this increment to the aec/bh cattle and there will result

$$(bdfh - ecah - bdcf + ecfa)/bh(f - c)$$

as the number of cattle which the meadow e suffices to pasture during time h. Proportionately, therefore, the meadow g will suffice to pasture

$$g(bdfh - ecah - bdcf + ecfa)/beh(f - c)$$

cattle during the same time h.

EXAMPLE. *If 12 cattle eat up $3\frac{1}{3}$ acres of meadow in 4 weeks and 21 cattle eat up 10 acres of exactly similar meadow in 9 weeks, how many cattle shall eat up 36 acres in 18 weeks?*
Answer: 36. To be sure that number will be found by substituting in

$$g(bdfh - ecah - bdcf + ecfa)/beh(f - c)$$

the numbers 12, $3\frac{1}{3}$, 4, 21, 10, 9, 36 and 18 for the letters a, b, c, d, e, f, g and h respectively. But the solution will perhaps be no less speedy if it be derived from first principles on the pattern of the preceding algebraic solution. Precisely, if 12 cattle in 4 weeks eat up $3\frac{1}{3}$ acres, then, proportionately, 36 cattle in 4 weeks, or 16 cattle in 9 weeks, or 8 cattle in 18 weeks should eat up 10 acres—if, that is, the grass were not to grow. But since, because of the growth of the grass, 21 cattle in 9 weeks eat up only 10 acres, that growth of grass in 10 acres during the latter 5 weeks will in itself be enough to pasture the excess of 21 over 16, that is, 5 cattle during 9 weeks, or, what is exactly the same, $\frac{5}{2}$ cattle during 18 weeks. And in 14 weeks (the excess of 18 over the first 4) the growth in the grass will proportionately be enough to pasture 7 cattle during 18 weeks; for 5 weeks : 14 weeks = $\frac{5}{2}$ cattle : 7 cattle. Consequently, to the 8 cattle which 10 acres can pasture without any growth of its grass add these 7 cattle which the growth of the grass by itself is sufficient to feed, and the total will be 15 cattle. Finally, if 10 acres suffice to pasture 15 cattle during 18 weeks, then, proportionately, 36 acres are sufficient during the same period for 54 cattle.

2.2. Arithmetical and combinatorial problems

We shall now turn to some other problems involving various properties of integers such as divisibility, greatest common factors, common multiples etc. or combinations, permutations, etc., etc. Such problems, "pleasant and delightful" Bachet de Méziriac calls them, have always been popular, and Alcuin* provides us with examples like the following:

> A wolf, a goat and a cabbage are on one bank of a river and the only boat is so small that the boatman can take no more than one of them across at a time. How are they all to get across the river without the wolf harming the goat or the goat harming the cabbage?

Here is a moral problem to be found in the *Entertainments* (*Récréations*) of Leurechon:

> There are fifteen Christians and fifteen Turks aboard a ship which runs into a storm. The captain says that the only way to save the vessel is to lighten her by throwing half of the passengers into the sea. The choice must clearly be made by lot, and the method agreed upon is that all the passengers shall stand in line, and the captain shall then count along the line in nines and throw every ninth passenger into the sea, until there are only fifteen passengers left. However, since he is himself a Christian, the captain wants to save the Christians: how can he arrange the passengers so that it is always a Turk and never a Christian that is in the ninth place?

Leurechon has merely copied this problem from Bachet. In fact, it can be found even earlier, in the work of Chuquet, though Chuquet divides the passengers into Christians and Jews. The traditional enemy varies over the centuries.

Alcuin provides a Western example of a very important type of problem which is also found in Hindu, Chinese and Arabic works dating from the Middle Ages:

> A hundred bushels of corn are shared out among a hundred people. Each man gets three bushels, each woman two bushels and each child half a bushel. How many men, women and children must there be?

The first Western mathematician to give a general and rigorous treatment of this kind of indeterminate problem was Bachet. There are seven possible solutions† to the above

---

* The collection of problems entitled *Propositiones ad acuendos juvenes* cannot be attributed to Alcuin with absolute certainty, but it undoubtedly dates from before 1000 AD.

† There are only six solutions—unless we admit the one where there are *no* women. J.V.F.

problem. Alcuin gives only one of them : 11 men, 15 women and 74 children.

Chuquet gives an example of a similar common problem :

A woman was carrying some eggs to market. On the way a man made her drop the eggs and had to pay her for them. But the woman did not know how many eggs she had had, though she did know that if she counted them two by two, three by three, four by four, five by five or six by six she always had one egg left over. And when she counted them seven by seven there were no eggs left over. How many eggs did the woman have?

PLATE V. Babylonian tablet, in the University of Yale collection, showing the square root of 2.

3. Second Degree Problems

I have added up the area and the side of my square: 45'.
(Babylonian tablet)

3.1. Square roots

This very old problem, like many others, is found in both square root, i.e., solving the equation $x^2 = a$.

This very old problem, like many others, is found in both a numerical and a geometrical form: we are either required to calculate the side of a square whose area is of given numerical size or to construct the side of a square whose area is given geometrically.

Let us consider the numerical problem first.

The Babylonians made use of a table of squares to obtain upper and lower bounds for the value of the square root.

On one tablet, for example, we find the following list (the numbers are sexagesimal):

Number	Square
....	....
1;20	1;46,40
1;21	1;49,21
1;22	1;52,04
1;23	1;54,49
1;24	1;57,36
1;25	2;00,25
....	....

We can see from this table that the square root of 2 lies between 1;24 and 1;25.

In most of their calculations the Babylonians in fact adopted the value 1;25.

However, as we can see from the Old-Babylonian tablet shown in Plate V, a much more accurate value was also known, namely 1;24,51,10. Complicated tables or very heavy calculations must have been required to obtain a value as accurate as this.

We know that at least one method of extracting square roots was known to Greek, Hindu, Arab, and Byzantine scholars and to Western scholars by the time of the thirteenth-century Renaissance. Heron explains this method in his *Metrica*, by using it to find the square root of 720:

Since 720 has not got a rational root we shall find a close approximation to its root as follows. The next square greater than 720 is 729, which has a root of 27. We divide 720 by 27. The result is $26\frac{2}{3}$. We add 27, making $53\frac{2}{3}$, which we divide by two, obtaining $26\frac{1}{2}\frac{1}{3}$. The root of 720 is thus very close to $26\frac{1}{2}\frac{1}{3}$. In fact, if we multiply $26\frac{1}{2}\frac{1}{3}$ by itself we obtain $720\frac{1}{36}$, a difference of $\frac{1}{36}$.

If we require a difference even smaller than $\frac{1}{36}$, we start with $720\frac{1}{36}$ instead of with 729 and if we proceed as before we shall find a difference very much smaller than $\frac{1}{36}$.

So, if a_1 is an approximate value of $\sqrt{A}$, Heron takes as a second approximation the value $a_2 = \frac{1}{2}(a_1 + A/a_1)$, and, if required, a third approximation $a_3 = \frac{1}{2}(a_2 + A/a_2)$. The process can clearly be continued indefinitely.

Nicolas Rhabdas employs a very similar method. To find the root of 3:

I find the nearest square. This is 4, which has a root of 2. Twice this root gives 4. Since the number 3 is one unit less than the square 4, we divide this difference of one by 4, which gives us $\frac{1}{4}$. We subtract this $\frac{1}{4}$ from the root of 4, that is from 2, and are left with $1\frac{1}{2}\frac{1}{4}$. This number is the root of 3, the number $1\frac{1}{2}\frac{1}{4}$. If we multiply it by itself it gives 3 units and $\frac{1}{4}$ of a quarter, that is, $\frac{1}{16}$ of a unit. [1]

The differences between the two methods are merely superficial. If $A = a_1{}^2 \pm r$, Heron's method gives

$$a_2 = \frac{1}{2}\left(a_1 + \frac{a_1{}^2 \pm r}{a_1}\right) = a_1 \pm \frac{r}{2a_1},$$

the value obtained by Rhabdas's method. Rhabdas continues:

Having employed a crude method of finding the root, we now turn to a more refined one. We proceed as follows: expressing $1\frac{1}{2}\frac{1}{4}$ in quarters we have $\frac{7}{4}$, expressing the 3 units in quarters we have $\frac{12}{4}$. We now divide 12 by 7, the quotient is $1\frac{1}{2}\frac{1}{7}\frac{1}{14}$. So if you multiply $1\frac{3}{4}$ by $1\frac{10}{14}$ you will obtain exactly 3, neither more nor less. As for the exact root, it is found as follows: We add the two roots we have found, that is $1\frac{1}{2}\frac{1}{4}$ and $1\frac{1}{2}\frac{1}{14}$. The sum is 3 units and $\frac{18}{28}$, which halved gives us $1\frac{1}{2}$ and $6\frac{1}{2}$ twenty-eighths. Because of the $\frac{1}{2}$ I double this last numerator and obtain 13, no longer twenty-eighths but now fifty-sixths. Thus in all we have $1\frac{41}{56}$, which is the exact root of 3. The product of $1\frac{41}{56}$ by itself is obtained as follows: once 1, 1; once $\frac{41}{56}$, $\frac{41}{56}$; once more the same 41, again 41, in all $\frac{82}{56}$; finally $\frac{41}{56}$ multiplied by itself, $\frac{1681}{56}$ fifty-sixths, that is $\frac{30}{56}$ and $\frac{1}{56}$ of a fifty-sixth. Now 30 and 82 make $\frac{112}{56}$, that is two units. Adding this to 1 we therefore obtain 3 and $\frac{1}{56}$ of a fifty-sixth, or $\frac{1}{3136}$ of a unit.

Although the second part of Rhabdas's calculation takes us back to Heron's method, most Renaissance workers in fact repeated the first part of his procedure and pointed out that it can be repeated an unlimited number of times.

Both methods lead to approximate values which are greater than the actual root.

Al-Karkhi gives a method linear interpolation which leads to an approximate value less than the actual root: if $A = a_1{}^2 + r$, he takes as his next approximation the value

$$a_2 = a_1 + \frac{r}{2a_1 + 1}.$$

The method used today to take the square root of a large number, by working through its successive figures, is in fact merely a variant of these ancient methods.

The Greeks used this method for numbers written in alphabetic notation, and it is described in various thirteenth-century treatises on algorithms, such as that of John Sacrobosco.

However, once the integral part of the root has been determined all the writers adopt one of the two methods described above, or some similar method, in order to improve upon their approximate value.

In 1572 Bombelli employed a procedure which led to continued fractions. He explains it as follows:

Let us suppose we are required to find the root of 13. The nearest square is 9, which has root 3. I let the approximate root of 13 be 3 plus 1 tanto.* Its square is 9 plus 6 tanti p. 1 power.† We set this equal to 13. Subtracting 9 from either side of the equation we are left with 4 equal to 6 tanti plus 1 power.

Many people have neglected the power and merely set 6 tanti equal to 4. The tanto then comes to $\frac{2}{3}$ and the approximate value of the root is $3\frac{2}{3}$ since it has been set equal to 3 p. 1 tanto. However, taking the power into account, if the tanto is equal to $\frac{2}{3}$, the power will be $\frac{2}{3}$ of a tanto,‡ which, added to the

* 1 tanto = 1x, the unknown.

† $9 + 6x + x^2$.

‡ If $x = \frac{2}{3}$, $x^2 = x \times x = \frac{2}{3}x$.

6 tanti, will give us $6\frac{2}{3}$ tanti, which are equal to 4. So the tanto will be equal to $\frac{3}{5}$, and since the approximate root is 3 p. 1 tanto it comes to $3\frac{3}{5}$. But if the tanto is equal to $\frac{3}{5}$, the power will be $\frac{3}{5}$ of a tanto and we obtain $6\frac{3}{5}$ tanti equal to 4 . . . .

This gives a new approximate value of $3\frac{20}{33}$ etc. . . .

Positional decimal notation made it easy to find the square root of a large number to the nearest integer, and approximate roots were soon found for many numbers which were not perfect squares.

For example, in the twelfth century John of Seville calculated $\sqrt{2}$ by taking

$$\sqrt{2} = \tfrac{1}{1000}\sqrt{2\,000\,000} \simeq \tfrac{1}{1000} \times 1414.$$

He then expressed his result in sexagesimal fractions and wrote $\sqrt{2}$ in the form: $1°24'50''24'''$.

His method is interesting, but his result is not as accurate as that obtained by the Babylonians since he did not carry his calculations sufficiently far.

In 1565 Pierre Forcadel made the following comments on the method used by John of Seville:

> In our third book we have shown how to find an approximate square root of a number which is not a perfect square, expressing this root in terms of any denominator. We multiply together the given number and the square of the denominator, when the square root of the product will be an approximation to the number of parts required. We also said that we commonly employ 10, 100 or 1,000 etc. as the denominator of the parts, that is, we usually want the approximate square root of a number expressed in tenths, hundredths or thousandth parts etc. because it is easier to square such numbers or units, also our numbers are then not made up of different sorts of number, and a number which is not a square is much more easy to multiply by one of the squares of these units than it would be to multiply by the squares of other numbers.

In 1556, Tartaglia, who uses al-Karkhi's method, complains that the other method, used by Oronce Finé, is extremely slow. In fact, this method was only generally adopted after Stevin had introduced the use of decimals.

It can, however, easily be combined with Heron's method.

Thus, starting from the approximate value $\sqrt{2} \simeq 1\cdot4142$ we can calculate, to eight decimal places, the quotient

$$2/1\cdot4142 \simeq 1\cdot141422712$$

and then take as the value of the root the mean value $1\cdot41421356$, which is correct to eight decimal places.

Constructing the square root geometrically is equivalent to constructing a geometric mean, and the procedure is described by Euclid in Book VI of the *Elements*.

3.2. Quadratic equations

Quadratic equations appear in Old-Babylonian work. The following examples are taken from *Textes mathématiques babyloniens* of Thureau-Dangin [1].*

> 1. I have added up the area and the side of my square: 45'. You write down 1, unity. You divide 1 into two parts: (30'). You multiply [30'] and 30': 15'. You add 15' to 45': 1. This is the square of 1. From 1 you subtract 30', by which you multiplied: 30', the side of the square.
>
> 2. I have subtracted the side of my square from its area: 14'30. You write down 1, unity. You divide 1 into two parts: (30'). You multiply 30' and 30': 15'. You add it to 14'30: 14'30'15'. This is the square of 29°30. You add 30', by which you multiplied, to 29°30: 30, the side of the square.
>
> 7. [I have added up seven times the side of my square and] eleven times [the area: 6°15'. You write down 7 and 11. You multiply] 6°15' by 11: [1'8°45'. You di]vide [7 into two parts]: (3°30'). [You multiply together] 3°30' and 3°30': [12°15']. You add this [to 1'8°45': 1'21. This is the square of 9. You subtract 3°30', by which you multipli]ed from 9: [you write 5°30'. The reciprocal of 1] 1 cannot be found. [By what must I multiply 11, to] obtain [5°30' ? 30', the quotient. The side of the squa]re is 30'.

The first problem is to solve the equation $x^2 + x = \frac{3}{4}$. The scribe takes the coefficient of x, 1, divides it by 2 which gives $\frac{1}{2}$, and squares this; obtaining $\frac{1}{4}$. He adds $\frac{1}{4} + \frac{3}{4} = 1$, which has root 1. He subtracts $\frac{1}{2}$ from this root, and obtains an answer of $\frac{1}{2}$.

---

* See footnote, p. 50.

In fact, he takes the root of an equation of the form $x^2 + px = q$ to be given by

$$x = \sqrt{\frac{p^2}{4} + q} - \frac{p}{2}.$$

In problem 2 the scribe solves the equation $x^2 - px = q$ by using the formula

$$x = \sqrt{\frac{p^2}{4} + q} + \frac{p}{2}.$$

Problem 7 concerns an equation of the form

$$ax^2 + bx = c.$$

The scribe calculates first ac, then

$$\frac{b}{2}, \frac{b^2}{4}, ac + \frac{b^2}{4}, \sqrt{ac + \frac{b^2}{4}} \quad \text{and} \quad \sqrt{ac + \frac{b^2}{4}} - \frac{b}{2}.$$

He then takes

$$x = \frac{\sqrt{ac + \frac{b^2}{4}} - \frac{b}{2}}{a}$$

as the root of the equation.

This particular tablet gives twenty-four such problems.

We shall now move forward in time and consider one of the examples in al-Khovarizmi's *Algebra*:

A power and roots* equal to a number, this is as if one were to say: A power and six roots are equal to thirty-nine drachmae,† which means that if we add together one power and ten times its root the sum is thirty-nine. We proceed by the following rule: take half the number of roots, in this case five. Multiplied by itself this gives twenty-five. Add thirty-nine to this, which gives sixty-four. Take the square root, which is eight. Take away five. We are left with three, which is the root of the power. The power is nine.

---

* Power: *census* in the Latin text of Gherardo of Cremona; root: *radix*.
† Drachma: a coin, here used as a unit.

Al-Khovarizmi investigates various types of quadratic equation, ensuring that the coefficients on both sides of the equation shall be positive. Moreover, in Gherardo's translation at least, all numbers are written out in full and no symbols are used.

Al-Khovarizmi considers the following cases:

$$x^2 + px = q \qquad (1)$$

and

$$ax^2 + bx = c,$$

which can be reduced to the previous form by dividing through by a (p, q, a, b and c are all positive, as are the coefficients in all the following equations);

$$x^2 + q = px \qquad (2)$$

and

$$ax^2 + c = bx.$$

The roots are:

$$\frac{p}{2} - \sqrt{\frac{p^2}{4} - q} \quad \text{or} \quad \frac{p}{2} + \sqrt{\frac{p^2}{4} - q}.$$

The problem is classified as insoluble if $q > p^2/4$. If $p^2/4 = q$, the root is $p/2$.

For $$px + q = x^2 \quad \text{or} \quad bx + c = ax^2 \qquad (3)$$

the root is given by

$$x = \sqrt{\frac{p^2}{4} + q} + \frac{p}{2}.$$

The rules are first stated in terms of numerical examples, as in the first case we quoted, and then justified by means of geometrical figures.

First case: A square whose area is the required power is surrounded by four rectangles whose length is the side of the square (the root) and whose width is $\frac{1}{4}$ of the coefficient of p ($2\frac{1}{2}$ in al-Khovarizmi's numerical example) (Fig. 3.1).

d	h	
		a
t	Power	g
b		
	k	e

Fig. 3.1

The four squares like d and e have side $p/4$ and their areas are therefore known. Their total area is $p^2/4$.

Now the given equation states that the central square and the four rectangles have a total area q. The area of the large square is therefore known, and so therefore is its side. This justifies the formula.

The author also uses the figure shown below to derive this formula in another way (Fig. 3.2). The square ab is

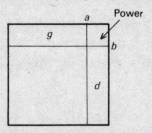

Fig. 3.2

the *power* and its side is the required root. The rectangles g and d are both of length $p/2$. The three areas make up a

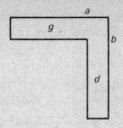

Fig. 3.3

gnomon (a Greek term for an "L" shaped figure) whose area is given by the equation as q (Fig. 3.3).

Now, taken together with the square of side $p/2$, the gnomon forms a large square with side $p/2 + x$ and known area $p^2/4 + q$. This gives us the formula.

Second case: The author considers a numerical example: $x^2 + 21 = 10x$.

A band of width x and length 10, and thus with area $10x$, is made up of the square x^2 and an area equal to 21 (Fig. 3.4).

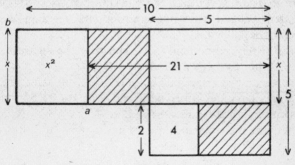

Fig. 3.4

A square of side 5 is made up of a gnomon of the same area as the previous one, 21, together with a small square, whose area is therefore $25 - 21 = 4$, and whose side is therefore 2. The width of the gnomon, x, is therefore $5 - 2 = 3$.

Al-Khovarizmi does not describe a graphical method of finding the other root of this equation, $x = 7$.

Third case:

$$3x + 4 = x^2.$$

The square (ad) is equal to the power, x^2, and its side to the root, x (Fig. 3.5).

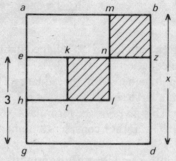

Fig. 3.5

On the side ga we take $eg = 3$. The rectangle (ed) thus has area $3x$, and, from the equation, the rectangle (az) therefore has area 4.

Let h be the mid-point of eg. We construct the square (hm) and the square (hk) with area $(1\frac{1}{2})^2 = \frac{9}{4}$.

The rectangles (mz) and (kl) both have the same area, since they have the same width: $mb = hg = kt = 1\frac{1}{2}$, and the same length: $tl = ae = bz = x - 3$.

Therefore the rectangle (az) = 4 has the same area as the gnomon *amltke*. The square (al), which is formed from the gnomon (area 4) and the square (hk) (area $\frac{9}{4}$), is therefore of known area $\frac{9}{4} + 4 = \frac{25}{4}$. Its side is known: $\frac{5}{2}$, so $x = 1\frac{1}{2} + \frac{5}{2} = 4$.

3.3. Euclid

The Old-Babylonian tablet we have discussed dates from a period about a thousand years earlier than that of Euclid, and the work of al-Khovarizmi dates from about a thousand years after him.

All of Euclid's geometry, which is concerned only with problems that can be solved by means of a straight edge and compasses, could be seen as concerned with quadratic problems or sets of quadratic problems, considered geometrically. The bases of his method are described in Book II of the *Elements*.

The book contains two definitions and fourteen propositions.

DEFINITION 1. Any rectangular parallelogram is said to be *contained* by the two straight lines containing the right angle.

DEFINITION 2. And in any parallelogram area let any one whatever of the parallelograms about its diameter with the two complements be called a *gnomon*.

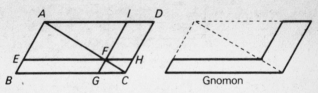

Fig. 3.6

The parallelogram *ABCD* has diameter (diagonal) *AC* (Fig. 3.6).

The parallelogram *GCHF* is said to be described about this diameter. Its complements are the parallelograms *BGFE* and *FHDI*. These three parallelograms combine to form the gnomon.

PROPOSITION 1. *If there be two straight lines, and one of them be cut into any number of segments whatever, the rectangle contained by the two straight lines is equal to the rectangles contained by the uncut straight line and each of the segments.*

This would be expressed algebraically as $a(b + c + d) = ab + ac + ad$. See Fig. 3.7.

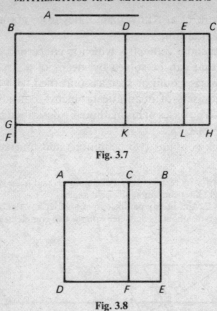

Fig. 3.7

Fig. 3.8

PROPOSITION 2. *If a straight line be cut at random, the rectangle contained by the whole and both of the segments is equal to the square on the whole.*

In algebraic terms: $(b + c)^2 = (b + c)b + (b + c)c$. See Figure 3.8.

PROPOSITION 3. *If a straight line be cut at random, the rectangle contained by the whole and one of the segments is equal to the rectangle contained by the segments and the square on the aforesaid segment.*

In algebraic terms: $(b + c)b = bc + b^2$. See Figure 3.9.

PROPOSITION 4. *If a straight line be cut at random, the square on the whole is equal to the squares on the segments and twice the rectangle contained by the segments.*

For let the straight line AB be cut at random at C; I say that the square on AB is equal to the squares on AC, CB and twice the rectangle contained by AC, CB.

For let the square $ADEB$ be described on AB [I, 46], let BD be joined; through C let CF be drawn parallel to either AD or EB, and through G let HK be drawn parallel to either AB or DE [I, 31].

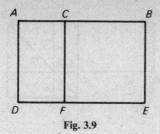

Fig. 3.9

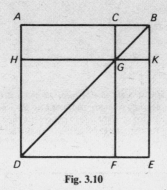

Fig. 3.10

Euclid shows (Fig. 3.10) that both *CGKB* and *HDFG* are squares. This proves the proposition.

In algebraic terms: $(a + b)^2 = a^2 + b^2 + 2ab$.

PROPOSITION 5. *If a straight line be cut with equal and unequal segments, the rectangle contained by the unequal segments of the whole together with the square on the straight line between the points of section is equal to the square on the half.*

In Figure 3.11, *C* is the mid-point of *AB* and *CBFE* is a square.

The area of the rectangle *ACLK* is equal to that of the rectangle *DBFG*.

Therefore the area of the rectangle *ADHK* is equal to that of the gnomon which Euclid marks with the arc *NOP*. It is clear that the sum of this gnomon and the square

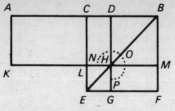

Fig. 3.11

$LHGE$ is the square $CBFE$, which was what we were required to prove.

In algebraic terms:

$$bc + \left(\frac{b-c}{2}\right)^2 = \left(\frac{b+c}{2}\right)^2.$$

PROPOSITION 6. *If a straight line be bisected and a straight line be added to it in a straight line, the rectangle contained by the whole with the added straight line and the added straight line together with the square on the half is equal to the square on the straight line made up of the half and the added straight line.*

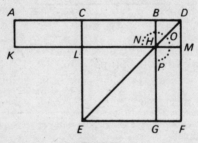

Fig. 3.12

In Figure 3.12, C is the mid-point of AB, and the line BD is added to AB in a straight line.

The rectangle $ADMK$ has length AD and width $DM = BD$, and its area is the same as that of the gnomon NOP. If we add to this gnomon the square LG, which has side CB, we obtain the square CF, which has side CD.

In algebraic terms:

$$c(c + b) + \frac{b^2}{4} = \left(\frac{b}{2} + c\right)^2.$$

PROPOSITION 7. *If a straight line be cut at random, the square on the whole and that on one of the segments both together are equal to twice the rectangle contained by the whole and the said segment and the square on the remaining segment.*

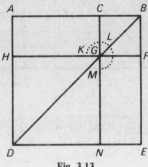

Fig. 3.13

Let C be a point on the line AB (Fig. 3.13). The square of AB is equal to the square HN, which has side AC, plus the gnomon KLM. The gnomon is equal to the rectangle AF, which has sides AB and $BF = CB$, plus the rectangle GE.

If we add the square CF to the gnomon we therefore obtain the rectangle AF plus the rectangle CE, i.e. twice the rectangle AF.

Therefore, altogether, square AE + square CF = square HN + twice the rectangle AF.

In algebraic terms: $(b + c)^2 + c^2 = 2(b + c)c + b^2$.

PROPOSITION 8. *If a straight line be cut at random, four times the rectangle contained by the whole and one of the segments together with the square on the remaining segment is equal to the square described on the whole and the aforesaid segment as on one straight line.*

We cut off the segment BC from the line AB, and take the point D such that $BD = BC$ (Fig. 3.14).

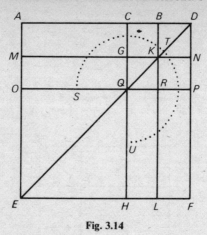

Fig. 3.14

The rectangle contained by AB and $BK(=CB)$ is AK, which is equivalent to MR or GL or KF.

Four times this rectangle gives the sum of the rectangles AR and GF, a sum which covers the square GR twice. Taking away one area of this square and transporting it to coincide with the square BN we obtain the gnomon STU. If we add the square OH to the gnomon we obtain the square AF.

In algebraic terms: $4(b + c)c + b^2 = (b + 2c)^2$.

PROPOSITION 9. *If a straight line be cut into equal and unequal segments, the squares on the unequal segments of the whole are double of the square on the half and of the square on the straight line between the points of section.*

Let C be the mid-point of AB and D some point on AB, for example a point between C and B (Fig. 3.15).

We construct $CE = CA$ perpendicular to AB. Then

$$AD^2 + DB^2 = AD^2 + DF^2 = AF^2 = AE^2 + EF^2$$

(since $\angle AEF$ is a right angle).

But $AE^2 = 2AC^2$ and $EF^2 = 2GF^2 = 2CD^2$.

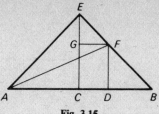

Fig. 3.15

In algebraic terms:

$$b^2 + c^2 = 2\left(\frac{b+c}{2}\right)^2 + 2\left(\frac{b-c}{2}\right)^2.$$

PROPOSITION 10. *If a straight line be bisected, and a straight line be added to it in a straight line, the square on the whole with the added straight line and the square on the added straight line both together are double of the square on the half and of the square described on the straight line made up of the half and the added straight line as on one straight line.*

Proof as before. $AD^2 + BD^2 = 2AC^2 + 2CD^2$ (see Fig. 3.16).

In algebraic terms:

$$(b + c)^2 + c^2 = 2\left(\frac{b}{2}\right)^2 + 2\left(\frac{b}{2} + c\right)^2.$$

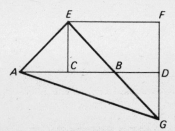

Fig. 3.16

PROPOSITION 11. *To cut a given straight line so that the rectangle contained by the whole and one of the segments is equal to the square on the remaining segment.**

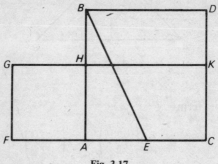

Fig. 3.17

Let *AB* be the given straight line [Fig. 3.17]; thus it is required to cut *AB* so that the rectangle contained by the whole and one of the segments is equal to the square on the remaining segment.

For let the square *ABDC* be described on *AB*; let *AC* be bisected at the point *E*, and let *BE* be joined; let *CA* be drawn through to *F*, and let *EF* be made equal to *BE*; let the square *FH* be described on *AF*, and let *GH* be drawn through to *K*.

I say that *AB* has been cut at *H* so as to make the rectangle contained by *AB*, *BH* equal to the square on *AH*.

In fact, by Proposition 6,

rectangle (*FK*) + square on *AE* = square on *EF*.

But, since *EF* = *EB*,

square on *EF* = square on *AE* + square on *AB*.

Therefore

rectangle (*FK*) = square on *AB* = square *AD*.

---

* This is the famous problem of dividing a line in extreme and mean ratio, the ratio known as the Golden Section. Euclid, *Elements*, Book VI, Definition 3: A straight line is said to have been *cut in extreme and mean ratio* when, as the whole line is to the greater segment, so is the greater to the less.

Subtracting the rectangle AK from both sides we have:

the square FH (which is the square on AH) is equal to the rectangle HD (contained by HB and $BD = AB$).

This is the first example of a geometrical solution to a quadratic equation. If we let $AB = 1$ and $AH = x$, the equation to be solved is

$$x^2 = 1 - x \quad \text{or} \quad x^2 + x = 1.$$

This problem is like the Babylonian ones discussed on p. 71. A Babylonian mathematician would have stated it in the form "I have added up the area and the side of my square: 1", and would have solved it by taking half the coefficient of x, that is $\frac{1}{2}$. Euclid takes $AE = AB/2$. A Babylonian would then have formed the square of half the coefficient: $\frac{1}{4}$, then added it to the constant 1, obtaining $\frac{5}{4}$, and finally he would have taken the square root of this number (which he would not have been able to find exactly).

Euclid adds up the squares, using Pythagoras's Theorem, and obtains EB^2, whose root is EB. The Babylonian would have subtracted $\frac{1}{2}$ from the square root. Euclid takes AF, which is the required solution.

The Babylonian scribe very often *checks* his numerical results.

Euclid *proves* that his solution is correct. In neither case is there any explanation of the method used to obtain the solution.

Euclid's geometrical constructions follow exactly the same pattern as the Babylonian calculations, but Euclid always arrives at an answer whereas the Babylonian calculations only give an exact answer if the square root can be found exactly. Euclid's method of geometrical proof was thus a great advance on the system of numerical checks which was used by the Babylonians.

Al-Khovarizmi combines the two techniques: he uses

numerical calculations in the manner of the Babylonians, and geometrical proofs in the manner of Euclid.

Propositions 12 and 13 of Book II of the *Elements* concern the values of the square of one side of a triangle when the opposite angle is acute and obtuse.

The final proposition of this book, Proposition 14, shows how to use a straight edge and compasses to find the side of a square of given area, so, taken together with Proposition 13 of Book VI, it thus describes the geometrical equivalent of calculating a square root.

A discussion of general methods of solving quadratic equations by geometrical means is to be found in Book VI of the *Elements*.

To apply an area to a straight line, or to form its *parabola*, is to construct on the given line a parallelogram with the given area.

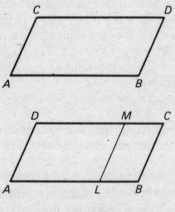

Fig. 3.18

Thus, in Figure 3.18 the parallelogram *ABDC* has been applied to *AB*.

The parallelogram *ALMD* has been applied to the line *AB* and is *in ellipsis* by the parallelogram *LBCM*.

To say that the parallelogram *ALMD* is "applied to the line *AB* in ellipsis by the parallelogram *LBCM*, or is deficient by *LBCM*" means that the parallelogram *LBCM* must be added on to the parallelogram *ALMD* if we are to form a parallelogram with base exactly equal to *AB*.

In problems the *ellipsis* or *defect LBCM* is of *given type*, i.e. it is to be similar to a given parallelogram.

In Proposition 27 of Book VI Euclid shows that the greatest of all the parallelograms which can be applied to the line *AB* deficient by a parallelogram similar to *OBQP* is the parallelogram constructed on half the line *AB*, i.e. on *AO* = *AB*/2 (Fig. 3.19).

This is an example of a problem which involves finding a maximum, a *diorism*. Its solution allows us to decide whether, in a given case, it is possible or impossible to apply a parallelogram in ellipsis.

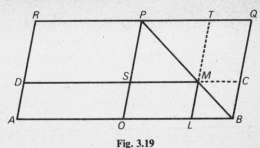

Fig. 3.19

The proof is straightforward. The parallelogram *ALMD* is equivalent to the gnomon *OBQTMS*, and its area is therefore less than that of the parallelogram *OBQP* or that of the equivalent parallelogram *AOPR*.

PROPOSITION 28. *To a given straight line apply a parallelogram equal to a given rectilineal figure and deficient by a parallelogrammic figure similar to a given one: thus the given rectilineal figure must not be greater than the parallelogram described on the half of the straight line and similar to the defect.*

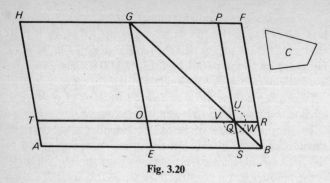

Fig. 3.20

It is required to apply to *AB* a parallelogram *ASQT* deficient by a parallelogram *SBRQ* similar to the given parallelogram *EBFG* (Fig. 3.20). The parallelogram *ASQT* is to be of given area *C*.

Since the parallelogram *OQPG* is equivalent to the gnomon *UWV*, and the area of the parallelogram *EBFG*, to which *OQPG* is similar, is known, the area of the parallelogram *OQPG* is also known.

Since the *type* (or shape) of this parallelogram is known, the length of its side *OQ* = *ES* can be deduced from its area. This gives us the point *S*, and enables us to construct the required parallelogram. It is clear that the problem only admits of a solution if the given area, *C*, is less than the area of the parallelogram *EBFG*. In algebraic terms: taking *AB* = *a*, we are required to find *AS* = *x*.

The areas of parallelograms of a given shape are proportional to the squares of their sides. Let the constant of proportionality be λ.

The parallelograms *EBFG* and *AEGH* are each of area $(\lambda/4)a^2$. The parallelogram *SBRQ*, the defect of *ASQT*, has side $(a - x)$ and area $\lambda(a - x)^2$. The parallelogram *ASQT*, which has the same height as *SBRQ* but has base *x*, is therefore of area:

$$\lambda(a - x)^2 \times \frac{x}{a - x} = \lambda(a - x)x.$$

Its area is given as c. Therefore we have

(1) $\lambda(a - x)x = c$

that is

(2) $\lambda x^2 - \lambda ax + c = 0.$

The form (2) was one of the standard forms used by the Babylonians.

Equation (1) poses the problem in the following terms: find two numbers x and $a - x$ such that their sum is a and their product is c/λ.

Thus, in geometrical terms, we are required to construct a rectangle whose semi-perimeter and area are given.

An Assyrian tablet of the Seleucid period, i.e. roughly contemporary with Euclid, considers a similar problem [3]:

I have added up the side and the front: 14. Also the area is 48.

Assuming you do not know, 14 times 14:3'16. 4 times 48:3'12. You subtract 3'12 from 3'16: the difference is 4. What must I multiply by what to obtain 4? 2 times 2:4. You subtract 2 from 14: the difference is 12. 30' times 12:6. The front is 6. You add 2 to 6:8. The side is 8.

Given that $x + y = a$ and $xy = b$, the scribe proceeds as follows:

He calculates a^2, $4b$, $a^2 - 4b$, $\sqrt{a^2 - 4b}$, $a - \sqrt{a^2 - 4b}$,

$$\frac{a - \sqrt{a^2 - 4b}}{2} = x \quad \text{and finally} \quad \frac{a + \sqrt{a^2 - 4b}}{2} = y.$$

Diophantus, who lived in Alexandria in the first century AD, also considers this problem in *Arithmetica* Book I, Proposition 27:

To find two numbers whose sum and product are given.

The problem is soluble only when half the sum of the required numbers is greater than the product of the numbers by a number which is a perfect square.

The condition is necessary only because Diophantus, like all Greek mathematicians, required that numbers should be either integers or fractions. There was no concept corresponding to what we should call an *irrational number*, and though reference was made to "lengths not commensurable with the unit of length" such quantities could be used only in geometry and not in arithmetic.

Diophantus solves the problem as follows:*

Let us suppose that the sum of the numbers is 20 units and that their product is 96 units.

Let the difference between the numbers be $2N$. Now, since the sum of the numbers is 20 units, if we divide it into two equal parts each part will be half the sum, that is 10 units. If we add to one of these parts half the difference between the numbers, that is $1N$, and if we subtract the same amount from the other part, we obtain two numbers whose sum is 20 units and whose difference is $2N$. So we may suppose that the greater of the required numbers is $1N$ plus 10 units, half the sum of the numbers, and the smaller number will then be 10 units minus $1N$, which makes the sum of the two numbers 20 units and their difference $2N$.

We also require that the product of the two numbers shall be 96 units. Now their product is 100 units $-$ $1Q$. This is equal to 96 units, so $1N$ is 2 units. Therefore the greater of the two numbers we require is 12 units and smaller is 8. These numbers satisfy the given conditions.

It will be noted that unlike Euclid, the Babylonians or al-Khovarizmi, Diophantus presents his solution in a form that allows us to follow his line of reasoning, instead of merely stating his result without proof.

Let us return to Euclid. Having considered application with defect, or *parabola in ellipsis*, Euclid next (Book VI, Proposition 29) turns to application with excess or *parabola in hyperbole*:

PROPOSITION 29. *To a given straight line to apply a parallelogram equal to a given rectilineal figure and exceeding by a parallelogrammic figure similar to a given one.*

---

* Diophantus has no symbol for addition but he does use a symbol for subtraction, which we have rendered by the modern $-$. He also uses a symbol for the square of the unknown, which we have rendered by Q.

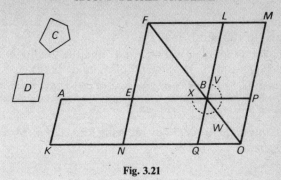

Fig. 3.21

It is required to apply to *AB* the parallelogram *APOK* in excess by a parallelogram *BPOQ* similar to a given parallelogram *D* (Fig. 3.21). It is also required that the area of the parallelogram to be applied shall be equal to a given area *C*.

On the line *EB* = *AB*/2 Euclid constructs the parallelogram (*EL*) similar to *D*.

Then on *FN* and *FM* he constructs a parallelogram *FNOM*, also similar to *D* and with area equal to the sum of *C* and the area of *EL*.

This construction gives him the required parallelogram *APOK* and he proves that it fulfils the required conditions by showing that it is equivalent to the gnomon *VWX*. Application *in hyperbole* can be expressed by the equation:

$$x(x - a) = b \quad \text{or} \quad x^2 - ax = b.$$

The first form of this equation expresses the problem as one of finding two numbers whose difference is *a* and whose product is *b*, a problem which occurs in Babylonian work, and in the work of Diophantus. The latter considers a numerical example in which *a* = 4 and *b* = 96.

He takes the larger number to be 2 + *N* (2 = *a*/2), and the smaller to be *N* − 2.

The product of the numbers, $Q - 4$, is equal to 96, so Q is equal to 100 and N is 10.

The larger number is therefore 12 and the smaller 8.

Euclid considers the same problem in his *Data*:

PROPOSITION 84. *If two straight lines contain a given area in a given angle and if one of them is longer than the other by the length of a given line, each of the lines is given.*

He solves the problem by reducing it to that of *parabola in hyperbole*.

PROPOSITION 85. *If two straight lines contain a given area, in a given angle, and if their sum is given, each of the lines is given.*

Euclid then appeals to *parabola in ellipsis*.

3.4. Applications to the conic sections

Archimedes calls the three conic sections: *section of a right cone*, *section of an acute cone* and *section of an obtuse cone*.

It was Apollonius who first gave them their modern names: parabola, ellipse and hyperbola.

Consider a cone with vertex S and circular base ENF (Fig. 3.22). Let G be the centre of the base. Any section BMC of the cone, taken parallel to the base, is a circle whose

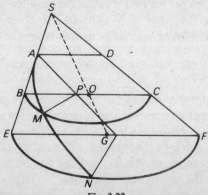

Fig. 3.22

centre, O, lies on the straight line SG. The plane BMC is variable, but it is always parallel to the base of the cone. Let $APMN$ be a fixed plane cutting the cone in the curve AMN. We shall investigate the properties of this curve.

The variable plane BMC cuts the plane APN in lines MP, whose direction remains fixed as the plane BMC moves.

We draw the line EGF, which is a diameter of the circular base of the cone and perpendicular to the direction of MP. Let BOC be the diameter of the circle BMC and parallel to the line EGF.

The line BOC always lies in the fixed plane ESF which cuts the plane AMP in a line AP which Apollonius calls the *diameter* of the conic section AMN.

In Apollonius's work the straight segment AP is given a name which is translated into Latin as *abscissa*, and the name given to the segment PM becomes the term *ordinate*. Generally, the ordinates make an oblique angle with the diameter.

Three cases must be considered.

FIRST CASE: When the diameter AP is parallel to one of the generators SB or SC. We shall assume AP is parallel to SC. The segment PC is therefore of constant length, and is equal to the segment AD, where AD is a line drawn parallel to EF.

In the circle BMC we have:

square on MP = rectangle $PB . PC$ = rectangle $PB . AD$.

But the triangles SAD and ABP are similar. Therefore $PB = AP (AD/SD)$, and the rectangle $PB . AD$ is equivalent to the rectangle with sides AP and L, where L is the constant length defined by $L = AD^2/SD$. We therefore have

square on MP = rectangle $AP . L$.

In other words, if we apply the square MP to the straight line L, we obtain the abscissa AP.

The exact *parabola* of the square of the ordinate on the *latus rectum* gives the abscissa.

The curve is called a *parabola*.

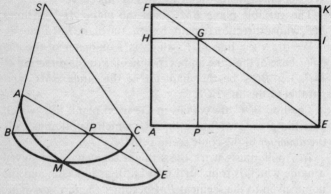

Fig. 3.23

SECOND CASE: When the diameter *AP* cuts the two lines *SB*, *SC* in the points *A* and *E* (Fig. 3.23).

An argument similar to the one employed above enables us to show that the square on *MP*, which is equal to the rectangle *PB . PC*, is proportional to the rectangle *PA . PE*.

If the constant of proportionality is the ratio of some length *AF* to the length *AE*, the square of the ordinate is equal to the rectangle *APGH*, that is, it is applied to the line *AE* in *ellipsis* by a rectangle *PEIG*, similar to the given rectangle *AEKF*.

The section is called an *ellipse*.

The length *AF* is the *latus rectum* with respect to the diameter *AE*.

THIRD CASE: When the diameter *AP*, produced, cuts *SB* at *A* and *SC*, produced, at *E* (Fig. 3.24).

In this case the square on *PM*, which is equal to the rectangle *PB . PC*, is proportional to the rectangle *PA . PE*.

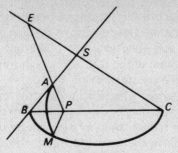

Fig. 3.24

Euclid would express this by saying that we had applied the square of the ordinate to the line *AE* in *hyperbole* by a rectangle similar to a given rectangle: hence the name *hyperbola* given to the curve.

3.5. Book X of Euclid's *Elements*

Book X is the longest of the books of the *Elements* and the most difficult one for an unprepared modern reader to understand. The book contains a hundred and fifteen propositions.

In Book X Euclid is mainly concerned with investigating the properties of such irrational numbers as arise from the taking of square roots, so it is appropriate to discuss the work at this point.

The first thirteen propositions are concerned with general properties of commensurable or incommensurable quantities. Euclid takes a straight line of arbitrary length *u* and calls it *the rational line*. We shall refer to this line as the *unit*. Any line is said to be *rational* if it is commensurable with the unit or if its square is commensurable with the square on the unit.*

The Greek term used is εὐθεῖα ῥητή, expressible line.

---

* Note that this is *not* the modern usage of the word "rational", which has, since the mid-nineteenth century, come to mean *expressible as a quotient of integers*. H.G.F.

We might describe this property as follows: the diagonal of a square is rational (expressible) if the side of the square is taken as the unit since the square of the diagonal is equal to twice the square of the side: *it is equal to two in square power* or 'it has a square power of 2'.

In modern terms, what Euclid calls the measure of a rational line would thus be expressed in terms of the chosen unit either by a rational number a (an integer or a fraction) or by a number in the form $\sqrt{a}$ where a is not a perfect square.

Euclid then describes as *rational* areas expressible in the form au^2. (In what follows we shall never use u to designate any rational number other than the unit.) An area is a *medial* if it is equal to $u^2\sqrt{a}$, i.e. if its measure is $\sqrt{a}$.

A straight line is a *binomial* if its measure is of the form $\sqrt{a} + \sqrt{b}$, where the ratio $a:b$ is not a perfect square. Two letters are always required to express such a number.

A straight line is called an *apotome* if its measure is of the form $\sqrt{a} - \sqrt{b}$, where the ratio $a:b$ is not a perfect square.

We shall not follow Euclid's own order of exposition but shall attempt instead to describe Book X in a manner which brings out those of its features which are still of importance to a mathematician of our own time.

The fundamental proposition is the one which states that the equation

$$\sqrt{a} = \sqrt{b} \pm \sqrt{c} \qquad (1)$$

is impossible if the ratio $b:c$ is not a square.

We can in fact deduce from (1) that

$$a = b + c \pm 2\sqrt{bc}$$

i.e.

$$\sqrt{bc} = r, \qquad \text{where } r = \tfrac{1}{2}|b + c - a|. \qquad (2)$$

If we write ck for r, which is always possible for a rational r, (2) gives $bc = c^2k^2$, i.e. $b = ck^2$, which contradicts the hypothesis that the ratio $b:c$ was not a square.

Having shown this we shall now consider the conditions for the square root of a binomial to be a binomial.

If

$$\sqrt{\sqrt{a} + \sqrt{b}} = \sqrt{c} + \sqrt{d}$$

we have $\sqrt{a} + \sqrt{b} = c + d + 2\sqrt{cd}$.

Therefore

$$a + b + 2\sqrt{ab} = (c + d)^2 + 4cd + 4(c + d)\sqrt{cd}.$$

Writing $|a + b - (c + d)^2 - 4cd|$ as $2r$ we obtain

$$\sqrt{ab} \pm r = 2(c + d)\sqrt{cd}.$$

From the preceding proposition we deduce that either ab is a perfect square, and $\sqrt{a} + \sqrt{b}$ is therefore not a true binomial ($b = ak^2$, $\sqrt{a} + \sqrt{b} = (1 + k)\sqrt{a}$), or r is equal to zero.

In the latter case $a + b = (c + d)^2 + 4cd$ and $\sqrt{ab} = 2(c + d)\sqrt{cd}$, i.e. $ab = 4(c + d)^2cd$.

If the sum and the product of two numbers are known then the two numbers themselves are known, so we may deduce from the above equations that $a = (c + d)^2$ and $b = 4cd$, which gives $a - b = (c - d)^2$.

The first term of the binomial is therefore rational, and the difference between the squares of the two terms is a perfect square: $p + \sqrt{p^2 - q^2}$.

Euclid calls such an expression a *first binomial*.

Similarly $p - \sqrt{p^2 - q^2}$ is a first apotome:*

$$\sqrt{p \pm \sqrt{p^2 - q^2}} = \sqrt{\frac{p + q}{2}} \pm \sqrt{\frac{p - q}{2}}$$

---

* The word *binomial* passed into Latin and eventually into English mathematical terminology, though in a wider sense, and gave rise to analogous words such as monomial, trinomial and polynomial. The word *apotome*, however, was translated in various ways, Campano used *residuum*, Zamberti used *apotome*, Montdoré returned to *residuum*, and Fibonacci used *recisorum* and Henrion used *résidu*. With the introduction of relative numbers (i.e. numbers preceded by a plus or minus sign) the term fell into disuse.

Since this case occurs very rarely, Euclid defines a new category of irrational lines: a *medial* line is one whose square has a medial area. We should express a medial as a fourth root: $\sqrt[4]{a}$.

A *bimedial** is the sum of two medials. Its measure is therefore $\sqrt[4]{a} + \sqrt[4]{b}$.

What is the condition that the square root of a binomial shall be a bimedial?

i.e. that
$$\sqrt{\sqrt{a} + \sqrt{b}} = \sqrt[4]{c} + \sqrt[4]{d}$$

It is not an easy matter to establish the necessary and sufficient conditions for such an equation to be possible.†

---

* Euclid in fact uses the word bimedial in a rather narrower sense which we shall describe later.

† Let $\sqrt[4]{c} = x$ i.e. $x^4 = c$, and $\sqrt[4]{d} = y$, i.e. $y^4 = d$, where x and y are both rational.

$$x + y = \sqrt{\sqrt{a} + \sqrt{b}}$$
$$\therefore \quad (x + y)^2 = \sqrt{a} + \sqrt{b}$$
$$\therefore \quad (x + y)^2 - \sqrt{b} = \sqrt{a}$$
$$\therefore \quad (x + y)^4 - 2\sqrt{b} \cdot (x + y)^2 + b = a$$
$$\therefore \quad [(x + y)^4 + b - a]^2 = 4b(x + y)^4.$$

We now expand this in terms of x and y, replacing x^4 and y^4 by their rational values whenever possible, e.g. we replace y by $x \times y/x$ and then replace x^4 by its rational value.

We obtain an equation of the form

$$M\left(\frac{y}{x}\right)^3 + N\left(\frac{y}{x}\right)^2 + P\left(\frac{y}{x}\right) + Q = 0.$$

Let $(y/x)^4 = d/c = r$. We may then write $\sqrt{r}$ for $(y/x)^2$, and our equation can be rewritten in the form

$$\frac{y}{x} = \frac{m\sqrt{r} + n}{m'\sqrt{r} + n'}.$$

(cont.)

Throughout this book Euclid's approach is synthetic. Since the square of $\sqrt[4]{c} + \sqrt[4]{d}$ is $\sqrt{c} + \sqrt{d} + 2\sqrt[4]{cd}$ he sets $\sqrt{c} + \sqrt{d} = \sqrt{a}$ and $2\sqrt[4]{cd} = \sqrt{b}$.

The first relation is only possible if $\sqrt{d} = k\sqrt{c}$, i.e. $d = k^2c$ (supposing $k < 1$). So

$$\sqrt{b} = 2\sqrt{kc}$$

$$\therefore \quad b = 4kc$$

and

$$a = (1 + k)^2c.$$

Euclid distinguishes between two possible cases: if kc is a perfect square $k = \lambda^2c$ and the binomial $\sqrt{a} + \sqrt{b}$ is a *second binomial*:

second binomial: $\sqrt{(1 + \lambda^2c^2)c} + 2\lambda c$

second apotome: $\sqrt{(1 + \lambda^2c^2)c} - 2\lambda c$

$$\sqrt{\sqrt{(1 + \lambda^2c)^2c} \pm 2\lambda c} = \sqrt[4]{c} \pm \sqrt[4]{\lambda^4c^3}.$$

Eliminating the root from the denominator we obtain

$$y = x(p\sqrt{r} + q).$$

But $(y/x)^2 = \sqrt{r}$, therefore, squaring both sides of the previous equation, we obtain

$$\sqrt{r} = p^2r + q^2 + 2p\sqrt{r},$$

i.e.

$$\sqrt{r} = \frac{p^2r + q^2}{1 - 2p}.$$

$\sqrt{r}$ is therefore rational. Let it be equal to s, say. Then $d/c = r = s^2$.

The bimedial $\sqrt[4]{c} + \sqrt[4]{d}$ can therefore be written in the form given by Euclid; namely $\sqrt[4]{c} + \sqrt[4]{cs^2}$.

Depending on whether the plus or the minus sign is taken the expression on the right is either a *first bimedial* or a *first apotome of a medial*.

If kc is not a perfect square, the binomial is a third binomial:

third binomial: $\sqrt{(1 + k)^2 c} + \sqrt{4kc}$

third apotome: $\sqrt{(1 + k)^2 c} - \sqrt{4kc}$

$$\sqrt{\sqrt{(1 + k)^2 c} \pm \sqrt{4kc}} = \sqrt[4]{c} \pm \sqrt[4]{k^2 c}.$$

If we take the plus sign, the expression on the right is a *second bimedial*, if we take the minus sign it is a *second apotome of a medial*.

Euclid continues his classification of binomials by considering those that do not satisfy the preceding conditions.

Assuming that $a^2 > b$ he describes as a *fourth binomial* the expression $a + \sqrt{b}$.

Its root, $\sqrt{a + \sqrt{b}}$, which cannot be simplified, is called a *major*.

$a - \sqrt{b}$ is a *fourth apotome*

$\sqrt{a - \sqrt{b}}$ is a *minor*.

If $a > b^2$

$\sqrt{a} + b$ is a *fifth binomial*,

and

$\sqrt{a} - b$ is a *fifth apotome*, and their roots are

$\sqrt{\sqrt{a} + b}$ which is the "*side*" *of the sum of a rational area and a medial area,*

and

$$\sqrt{\sqrt{a} - b}$$ which is the "*side*" *of an area which with a rational area produces a medial whole.*

Finally we have the general case: a binomial $\sqrt{a} + \sqrt{b}$ which satisfies none of the above conditions is called a *sixth binomial*, and $\sqrt{a} - \sqrt{b}$ is called a *sixth apotome*.

$$\sqrt{\sqrt{a} + \sqrt{b}}$$ is the "*side*" *of the sum of two medial areas*

$$\sqrt{\sqrt{a} - \sqrt{b}}$$ is the "*side*" *of an area which with a medial area produces a medial whole.*

Euclid's work is exceedingly thorough. He establishes the existence of each of the twelve large classes of straight lines, and that of each of the six classes of binomials as well as of the six classes of apotomes which give rise to them.

He carefully proves that his classes do not overlap. For example, an apotome cannot also be a binomial. For, if we have

$$\sqrt{a} - \sqrt{b} = \sqrt{c} + \sqrt{d}$$

then we have

$$a + b - 2\sqrt{ab} = c + d + 2\sqrt{cd}$$

and

$$a + b - c - d = 2\sqrt{ab} + 2\sqrt{cd}.$$

and this last equality cannot possibly hold, since the expression on the right is a non-zero binomial and the expression on the left is a rational number.

The following elementary results are also proved:

$$\frac{a}{\sqrt{b}} = \frac{a}{b}\sqrt{b},$$

and

$$\frac{a}{\sqrt{b} \pm \sqrt{c}} = \frac{a}{b - c}(\sqrt{b} \pm \sqrt{c}).$$

Euclid uses no symbols and he derives his results by using geometrical methods to solve quadratic equations: parabola, parabola in ellipsis by a square, parabola in hyperbola by a square. It is said that Apollonius wrote a work, now completely lost, in which he generalized Euclid's results. Since Euclid had classified and ordered certain kinds of irrational numbers Apollonius must have investigated the properties of those that remained unclassified.

The Greeks, at least in the mathematical works that have come down to us, never identified a non-rational square root with a number. The same naturally applied to combinations of such roots. In medieval times, however, the Hindus and then the Arabs did identify square or even cube roots with numbers.

Western scholars, led by Leonardo of Pisa, followed their example.

A whole branch of arithmetic, very like algebra, was to develop out of this work: namely the set of rules for handling *surds* or irrational numbers.

In the sixteenth century these rules were generally to be found in treatises on Algebra, and they explained how to write the sums, differences, products and quotients of square and cube roots.

At their highest level, such rules derive from Book X of Euclid's *Elements*. The work was in fact much studied for the purpose of finding such rules, and Arab and Western scholars repeatedly wrote commentaries on it from the thirteenth until the seventeenth centuries.

Leonardo of Pisa, among others, was well acquainted with Euclid's work.* Cardan devoted several chapters of his *Ars Magna* to Euclid's irrational numbers, making a detailed study of the equations which can have such numbers as their roots. In the fifteenth century Regiomontanus had also written a commentary on Book X. Tartaglia gave it an important place in his *General Trattato*, and Stevin devoted a special treatise to it, his *Appendix of incommensurable quantities* (*Appendice des incommensurables grandeurs*).

---

* When Leonardo wants to solve the equation

$$x^3 + 2x^2 + 10x = 20 \qquad (1)$$

he notes, before giving an approximate root, that the root of the equation cannot be one of Euclid's irrational numbers. [The equation has only one positive root.]

It is easy to verify this assertion. The equation can in fact be written in the form

$$x(10 + x^2) = 2(10 - x^2).$$

Squaring both sides, we obtain

$$x^2(10 + x^2)^2 = 4(10 - x^2)^2. \qquad (2)$$

Since the square of any of Euclid's irrational numbers is a binomial or an apotome let us write

$$x^2 = \sqrt{u} + \varepsilon\sqrt{v} \qquad (\varepsilon = \pm 1)$$

Equation (2) then gives us, after some manipulation,

$$\sqrt{u}(180 + u + 3v) + \varepsilon\sqrt{v}(180 + 3u + v) = 400 - 16u - 32\varepsilon\sqrt{uv}$$

For $\varepsilon = +1$ the left hand side is a binomial. The right hand side (which is positive since it is equal to the expression on the left) is an apotome or a rational number. This is impossible.

For $\varepsilon = -1$ we are again led to an impossible equation. The final stage of the proof depends upon the mutual exclusiveness of Euclid's classes of irrationals.

It was probably through studying Euclid's work that the Italian algebraists were led to invent algebraic methods of solving cubic and quartic equations.

Later on Book X was superseded by the work of other scholars and was forgotten.

Its spirit is, however, very like that of the works of the eighteenth- and early nineteenth-century algebraists who tried to solve equations in terms of square and other roots. The work of Ruffini, Abel, and Galois eventually proved that it was not possible to express the solution of the general quintic equation in such terms.

And later developments of this work eventually led to the idea of fields of algebraic numbers.

Book X of Euclid's *Elements* gave the original impetus to such work, though little trace of its influence now remains.

4. Pythagoras's Theorem

> *A door. Height: half a Ninda and 2 cubits.*
> *Width: 2 cubits. What is its diagonal?*
> Babylonian tablet

Euclid, Book I, PROPOSITION 47:

Ἐν τοῖς ὀρθογωνίοις τριγώνοις τὸ ἀπὸ τῆς τὴν ὀρθὴν γωνίαν ὑποτεινούσης
πλευρᾶς τετράγωνον ἴσον ἐστὶ τοῖς ἀπὸ τῶν τὴν ορθὴν γώνίαν περιεχουσῶν
πλευρῶν τετραγώνοις.

In right-angled triangles the square on the side subtending the right angle
is equal to the squares on the sides containing the right angle.

Henrion, in his translation of 1615, comments: "Now it is said that this celebrated and very famous theorem was discovered by Pythagoras, who was so full of joy at his discovery that, as some say, he showed his gratitude to the gods by sacrificing a Hecatomb of oxen. Others say he only sacrificed one ox, which is more likely than that he sacrificed a hundred, since this Philosopher was very scrupulous about shedding the blood of animals."

This theorem is attributed to Pythagoras only in late Greek works, and the attribution is sometimes tentative. Proclus, for instance, says "If we listen to those who like to record antiquities, we shall find them attributing this theorem to Pythagoras and saying that he sacrificed an ox on its discovery."[1]

It is now recognized that this famous theorem was in fact known considerably before the time of Pythagoras and was probably the earliest of all geometrical theorems.

The theorem can take three forms: the geometrical one, as stated by Euclid; a practical one, enabling an approximate value of the length of the last side of a triangle to be calculated from the lengths of the other two; and, lastly, an arithmetico-geometric one, in enabling us to construct right-angled triangles whose sides are integral multiples of the same unit. Such triangles are known as *Pythagorean triangles* and the simplest has sides 3, 4 and 5.

It is interesting that the earliest surviving references to this theorem are concerned with the last two forms. The following two examples are taken from Babylonian tablets.

The first example is from the Old-Babylonian period.

A palû, 30′, or a cane, in ... In height, it comes down from 6′, at the bottom how far away is it?

You, square 30′, you will get 15′. Subtract 6′ from 30′, you will get 24′. Square 24′, you will get 9′36″. Subtract 9′36″ from 15′, you will get 5′24″. Of what is 5′24″ the square? It is the square of 18′: it is 18′ away on the ground. If it is 18′ away on the ground what height did it come down from? Square 18′, you will get 5′24″. Subtract 5′24″ from 15′, you will get 9′36″. Of what is 9′36″ the square? It is the square of 24′. Subtract 24′ from 30′, you will get 6′; it came down from 6′. This is the way to proceed.

The second tablet is from the Seleucid period:

A reed is placed vertically against a wall. If it comes down from 3 cubits it is 9 cubits away at the bottom. What [length]* is the reed? What [height] is the wall?

Assuming you do not know, 3 times 3:9; 9 times 9:1′21. You add 9 to 1′21:1′30. You multiply 1′30 by 30′:45. The reciprocal of 3 is 20′. You multiply 20′ by 45: 15, [the length of] the reed. What [height] is the wall? 15 times 15: 3′45. 9 times 9: 1′21. You subtract 1′21 from 3′45: there remains 2′24. What must I multiply by itself to get 2′24? 12 times 12:2′24. [The height of] the wall is 12.†

---

* The words in square brackets [ ] are missing in the French and are presumably missing on the tablet. J.V.F.

† This can be compared with a problem given by Leonardo of Pisa (1202): Est hasta justa quamdam Turrim erecta habens in longitudinem pedes 20. Quare si pes hastae separatur a Turri pedibus 12. Quot pedibus caput hastae descenderit.

There are also several examples in which the length of an arrow is calculated from the radius of the bow and the length of the string.

But the most important of the ancient Babylonian mathematical texts is the one published in 1945 by Neugebauer and Sachs, which gives a list of Pythagorean triangles (see Plate VI, p. 109).

We notice that the numbers we have called b and d (in columns II and III) are, respectively, the base and the hypotenuse of a right-angled triangle whose height, h, given by $h^2 = d^2 - b^2$, can be expressed exactly in sexagesimal numbers in every case. The tablet is incomplete (a piece is missing on the left), so it does not give the heights, h, but the numbers in the first column are equal to d^2/h^2, so there is no doubt that we have here a table of Pythagorean triangles. As can be seen from the fairly regular decrease of the numbers in the first column, the triangles are classified in order of increasing angle opposite the height. This angle goes from about 45° for triangle 1 to about 60° for triangle 15.

The question inevitably arises as to how such a list, which shows a high degree of skill in computation, could have been compiled. Pythagoras's Theorem provides a possible method. Let us try to find Pythagorean triangles for which $h = 1$. We have $d^2 - b^2 = (d + b)(d - b) = 1$. If we let $d + b$ be equal to some number n then $d - b$ must be its reciprocal $1/n$. So a Babylonian mathematician would choose n to be a *regular* number (see Chapter 1), that is, a number whose reciprocal can be expressed exactly in the sexagesimal system. Then we shall have

$$d + b = n \quad \text{and} \quad d - b = \frac{1}{n},$$

which gives us

$$d = \frac{1}{2}\left(n + \frac{1}{n}\right) \quad \text{and} \quad b = \frac{1}{2}\left(n - \frac{1}{n}\right).$$

This is how historians explain the process by which the table was compiled, since new values of h, b and d can then be obtained by multiplying the previous values by various factors.

However, in this we are on uncertain ground. What can and should be said is the following:

We do not know of any Babylonian proof of Pythagoras's theorem.

We know that the Babylonians continually used this theorem in practical calculations, and we know that they knew how to generate any number of Pythagorean triangles.

We do not at present have any documents that indicate whether or not the Egyptians were acquainted with Pythagoras's theorem, but documents do survive from the early period of Hindu mathematics, in the fifth, fourth and third centuries BC.

The *Sulvasutras of Apastamba* describe a method of constructing right angles with a cord. To construct the perpendicular to Ox at O we hold a cord so that the parts OP, OQ and QP have as their respective lengths the three sides b, h and d of a Pythagorean triangle (Fig. 4.1).

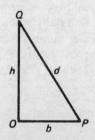

Fig. 4.1

PLATE VI. Babylonian tablet, Plimpton 322.

I	II (= b)	III (= d)	IV
[1, 59, 0,]* 15	1, 59	2, 49	1
[1, 56, 56,] 58, 14, 50, 6, 15	56, 7	3, 12, 1	2
[1, 55, 7,] 41, 15, 33, 45	1, 16, 41	1, 50, 49	3
[1,] 5[3, 10], 29, 32, 52, 16	3, 31, 49	5, 9, 1	4
[1,] 48, 54, 1, 40	1, 5	1, 37	5
[1,] 47, 6, 41, 40	5, 19	8, 1	6
[1,] 43, 11, 56, 28, 26, 40	38, 11	59, 1	7
[1,] 41, 33, 59, 3, 45	13, 19	20, 49	8
[1,] 38, 33, 36, 36	9, 1	12, 49	9
1, 35, 10, 2, 28, 27, 24, 26, 40	1, 22, 41	2, 16, 1	10
1, 33, 45	45	1, 15	11
1, 29, 21, 54, 2, 15	27, 59	48, 49	12
[1,] 27, 0†, 3, 45	7, 12, 1	4, 49	13
1, 25, 48, 51, 35, 6, 40	29, 31	53, 49	14
[1,] 23, 13, 46, 40	56	53	15

* The numbers enclosed in [ ] have been reconstructed by the editors. All the numbers are sexagesimal, and there are some errors on the tablet.

† On the tablet the 0 is indicated by a large gap between the figures.

Apastamba does not describe this procedure in a general manner but instead uses the following numerical examples:

h	b	d
3	4	5
12	16	20
15	20	25
5	12	13
15	36	39
8	15	17
12	35	37

We shall compare this list with what would be obtained by the Babylonian method of computation. This will give us some idea how Apastamba might have obtained his list, assuming that he, like his near contemporary Pythagoras, did not in fact discover the theorem independently.

As we have seen, the Babylonian method of computation leads to values of h, b and d which are respectively multiples of 2, $n - 1/n$ and $n + 1/n$. So $n = (b + d)/h$. Since b and h are used in the same way in the computation,* Apastamba's list gives values of n as follows: first of all 3, a Babylonian regular number ($\frac{1}{3} = 0; 20$), and then 2, a Babylonian regular number ($\frac{1}{2} = 0; 30$). These come from the first three triangles.

The fourth and fifth triangles give us:

$$n = 5, \qquad \text{a regular number } (\tfrac{1}{5} = 0; 12)$$

and

$$n = \tfrac{18}{12} = \tfrac{3}{2} \text{ the number } 1; 30, \text{ which is regular}$$

$$(\text{reciprocal } 0; 40).$$

---

* Between two numbers n and p which give the same triangle there exists the relation

$$p = \frac{n + 1}{n - 1} \quad \text{or} \quad np - n - p = 1.$$

For the sixth triangle: $n = \frac{25}{15} = \frac{5}{3}$, a regular number 1 ; 40 (reciprocal 0 ; 36), or $n = \frac{32}{8} = 4$, also a regular number (reciprocal 0 ; 15).

For the seventh triangle: $n = \frac{49}{35} = \frac{7}{5}$ an *irregular* number whose reciprocal cannot be expressed exactly, but also $n = \frac{72}{12} = 6$, a regular number (reciprocal 0 ; 10).

To sum up: all the triangles in Apastamba's list could be of Babylonian origin.

But Apastamba does not merely give a list of triangles. He also gives a general statement of Pythagoras's theorem: "*The diagonal of a rectangle produces the sum of what the largest and the smallest side produce separately.*"

Apastamba was also familiar with the result that as a special case of this theorem, the diagonal of a square is the side of a square with twice the area of the original one.

Apastamba can express an approximate value of this diagonal in terms of the sides:

$$1 + \frac{1}{3} + \frac{1}{12} - \frac{1}{12 \times 34}.$$

He constructs $\sqrt{3}$ as the diagonal of a rectangle with sides 1 and $\sqrt{2}$, and he also knows how to solve the general problem of constructing a square which is the sum or difference of two others.

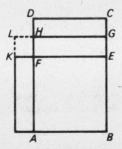

Fig. 4.2

His method of converting a rectangle into an equivalent square is as follows:

Let the given rectangle be *ADCB* (Fig. 4.2). We cut off from the long side *AD* a length *AF* equal to *AB*, the small side and form the square *AFEB*. We join the points *H* and *G*, the mid-points of the lines *DF* and *CE*, and transport the rectangle *HDCG* to the position (*AK*).

The area of the rectangle can thus be seen as the difference between the areas of the two squares *LF* and *LB*. Its side can be obtained as the second side of a rectangle whose width is *HF* and whose diagonal is *LG*.

Apastamba appears to be unable to solve the opposite problem: that of constructing a rectangle with given base *b* and area equal to that of a square of side *a*. He does not give a precise construction but merely says that we must subtract *ab* from a^2 and apply the remainder to the side *b*, which is a problem as difficult as the original one.

Apastamba's work is on about the same mathematical level as that of the Old-Babylonian texts and the approximately contemporary Greek work of the time of Pythagoras, before it was discovered that certain lengths were incommensurable with one another.

Pythagoras's theorem appears as Proposition 47 of Book I of the *Elements* and Euclid proves it as follows (Fig. 4.3):

Let *ABC* be a right-angled triangle having the angle *BAC* right; I say that the square on *BC* is equal to the squares on *BA*, *AC*.

For let there be described on *BC* the square *BDEC*, and on *BA*, *AC* the squares *GB*, *HC*; through *A* let *AL* be drawn parallel to either *BD* or *CE*, and let *AD*, *FC* be joined.

Then, since each of the angles *BAC*, *BAG* is right, it follows that with a straight line *BA*, and at the point *A* on it, the two straight lines *AC*, *AG* not lying on the same side make the adjacent angles equal to two right angles;

therefore *CA* is in a straight line with *AG*.

For the same reason

BA is also in a straight line with *AH*.

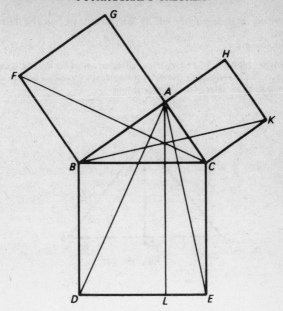

Fig. 4.3

And, since the angle *DBC* is equal to the angle *FBA*: for each is right:
let the angle *ABC* be added to each;
therefore the whole angle *DBA* is equal to the whole angle *FBC*.
And, since *DB* is equal to *BC*, and *FB* to *BA*, the two sides *AB*, *BD* are
equal to the two sides *FB*, *BC* respectively,
and the angle *ABD* is equal to the angle *FBC*;
therefore the base *AD* is equal to the base *FC*,
and the triangle *ABD* is equal to the triangle *FBC*.
Now the parallelogram *BL* is double of the triangle *ABD*, for they have
the same base *BD* and are in the same parallels *BD*, *AL*.
And the square *GB* is double of the triangle *FBC*, for they again have
the same base *FB* and are in the same parallels *FB*, *GC*.
[But the doubles of equals are equal to one another.]
    Therefore the parallelogram *BL* is also equal to the square *GB*.
    Similarly, if *AE*, *BK* be joined,
the parallelogram *CL* can also be proved equal to the square *HC*;
    therefore the whole square *BDEC* is equal to the two squares *GB*, *HC*.
    And the square *BDEC* is described on *BC*,
        and the squares *GB*, *HC* on *BA*, *AC*.

Therefore the square on the side *BC* is equal to the squares on the sides *BA*, *AC*.

Therefore etc.   Q. E. D.

PROPOSITION 48. *If in a triangle the square on one of the sides be equal to the squares on the remaining two sides of the triangle, the angle contained by the remaining two sides of the triangle is right.*

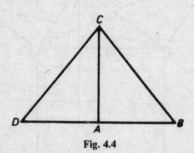

Fig. 4.4

For in the triangle *ABC* [Fig. 4.4] let the square on one side *BC* be equal to the squares on the sides *BA*, *AC*;

I say that the angle *BAC* is right.

For let *AD* be drawn from the point *A* at right angles to the straight line *AC*, let *AD* be made equal to *BA*, and let *DC* be joined.

Since *DA* is equal to *AB*, the square on *DA* is also equal to the square on *AB*.

Let the square on *AC* be added to each;

therefore the squares on *DA*, *AC* are equal to the squares on *BA*, *AC*.

But the square on *DC* is equal to the squares on *DA*, *AC*, for the angle *DAC* is right;

and the square on *BC* is equal to the squares on *BA*, *AC*, for this is the hypothesis;

therefore the square on *DC* is equal to the square on *BC*, so that the side *DC* is also equal to *BC*.

And, since *DA* is equal to *AB*,

and *AC* is common,

the two sides *DA*, *AC* are equal to the two sides *BA*, *AC*;

and the base *DC* is equal to the base *BC*;

therefore the angle *DAC* is equal to the angle *BAC*.

But the angle *DAC* is right;

therefore the angle *BAC* is also right.

Therefore etc.   Q. E. D.

B.N.

PLATE VII. Pythagoras's Theorem in Hérigone's *Cours de Mathématique* (1639).

It is interesting to compare Euclid's proof of Proposition 47 with that given by Hérigone, in 1639. Hérigone's proof follows Euclid at every stage, but uses a completely original system of notation. Equality is indicated by 2/2 (see Plate VII).

We shall now turn to less familiar material: the history of Pythagorean triangles. It was well known in ancient times that the diagonal of a square was incommensurable with its side. The principles from which this fact can be proved are mentioned by Aristotle, and we have reason to suppose that the result had in fact been proved considerably earlier, probably in the fifth century BC. The proof could have originated among the Pythagoreans. In Book X of the *Elements*, in a passage which seems to be an interpolation on the part of a commentator, the result is proved as follows:

We are required to prove that the diagonal of a square is incommensurable in length with its side.

Let *ABCD* be the square and *AC* its diagonal [Fig. 4.5]. I say that the straight line *AC* is incommensurable in length with *AB*.

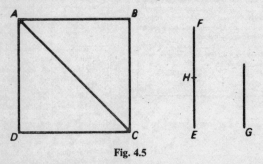

Fig. 4.5

Let us suppose that it is commensurable. I say that from this it follows that a number can at once be both odd and even.

For it is obvious that the square of *AC* is twice the square of *AB*. But *AC* is commensurable with *AB*. The ratio of the straight line *AC* to the straight line *AB* is therefore that of a number to a number.

Let the ratio of *AC* to *AB* be that of the number *EF* to the number *G* and let the numbers *EF* and *G* be the smallest numbers which bear this ratio to one another.

The number *EF* will not be unity.

For if *EF* were unity, since the ratio of *EF* to *G* is the same as the ratio of *AC* to *AB* and *AC* is greater than *AB*, *EF*, which is unity, would be greater than the number *G*, which is absurd. Therefore *EF* is not unity. Therefore *EF* is a number.

And since the ratio of *CA* to *AB* is the same as the ratio of *EF* to *G*, the ratio of the square of *CA* to the square of *AB* will be the same as the ratio of the square of *EF* to the square of *G*.

But the square of *CA* is twice the square of *AB*. The square of *EF* is therefore twice the square of *G*. The square of the number *EF* is therefore even.

The number EF is therefore even, since it if were odd its square would be odd. For if we add up any odd number of odd numbers their sum will be an odd number. The number EF is therefore even.

Let us divide the number EF into two equal parts at the point H. Since the numbers EF and G are the smallest which bear the given ratio to one another these numbers must be prime to one another. But the number EF is even. Therefore the number G is odd. For if it were even the numbers EF and G, which are prime to one another, would both be measured by two; because every even number has a part which is one half of it. This is impossible. The number G is therefore not even. Therefore it is odd.

But EF is twice EH. The square of EF is therefore four times the square of EH. But the square of EF is twice the square of G. The square of G is therefore twice the square of EH. The square of G is therefore even. The number G is therefore even, for the reasons already given.

But the number G is also odd, which is impossible. The straight line AC is therefore not commensurable in length with AB. It is therefore incommensurable with it.   Q. E. D.

Proclus alludes to this proof in his commentary on Proposition 47 of Book I of the *Elements,* remarking:[2]

There are two sorts of right-angled triangles, isosceles and scalene. In isosceles triangles you cannot find numbers that fit the sides; for there is no square number that is the double of a square number, if you ignore approximations, such as the square of seven which lacks one of being double the square of five. But in scalene triangles it is possible to find such numbers, and it has been clearly shown that the square on the side subtending the right angle may be equal to the squares on the sides containing it. Such is the triangle in the *Republic,** in which sides of three and four contain the right angle and five subtends it, so that the square on five is equal to the squares on those sides. For this is twenty-five, and of those the square of three is nine and that of four sixteen. The statement, then, is clear for numbers.

Certain methods have been handed down for finding such triangles, one of them attributed to Plato, the other to Pythagoras. The method of Pythagoras begins with odd numbers, positing a given odd number as being the lesser of the two sides containing the angle, taking its square, subtracting one from it, and positing half of the remainder as the greater of the sides about the right angle; then adding one to this, it gets the remaining side, the one subtending the angle.† For example, it takes three, squares it, subtracts one from nine, takes the half of eight, namely, four, then adds

---

* A dialogue by Plato.

† For an odd value of n we form the triangle with sides $n, (n^2 - 1)/2$ and $(n^2 + 1)/2$. We thus obtain all the Pythagorean triangles whose hypotenuse and second longest side are consecutive integers, and no other Pythagorean triangles.

one to this and gets five; and thus is found the right-angled triangle with sides of three, four, and five. The Platonic method proceeds from even numbers. It takes a given even number as one of the sides about the right angle, divides it into two and squares the half, then by adding one to the square gets the subtending side, and by subtracting one from the square gets the other side about the right angle.* For example, it takes four, halves it and squares the half, namely, two, getting four; then subtracting one it gets three and adding one gets five, and thus it has constructed the same triangle that was reached by the other method. For the square of this number is equal to the square of three and the square of four taken together.

It can be seen that the two methods attributed to Plato and to Pythagoras do not give us all the possible Pythagorean triangles. A more general method was used by the Babylonians, but their system of numerals and their system of computation compelled them to impose certain restrictions. The method does not require that n be an integer, but merely that n shall be rational. However, the Babylonian system of computation required that n should be regular. On the other hand, since $n = (b + d)/h$, there are two rational numbers $n = (b + d)/h$ and $p = (h + d)/b$ which correspond to each Pythagorean triangle.

The method is thus completely general. We shall express the fact that n is rational in an explicit form by writing $n = r/s$, where r and s are integers. Then

$$h = 2, \qquad b = \frac{1}{2}\left(\frac{r}{s} - \frac{s}{r}\right) \quad \text{and} \quad d = \frac{1}{2}\left(\frac{r}{s} + \frac{s}{r}\right).$$

The sides of Pythagorean triangles must be proportional to these values, so we may write them in the form $h = 2krs$, $b = k(r^2 - s^2)$, $d = k(r^2 + s^2)$.

This result is given in Book X of Euclid's *Elements*.

Diophantus's *Arithmetica* contains many problems which involve Pythagorean triangles. Vieta followed Diophantus's example, and in his *Notae Priores* Proposition 45

---

* If $2p$ is an even number we form the triangle $2p$, $p^2 - 1$, $p^2 + 1$. This gives us all the Pythagorean triangles in which the hypotenuse is two units longer than the next longest side, and no other Pythagorean triangles.

sets about forming a triangle in the following manner:

GENERATING TRIANGLES

PROPOSITION 45. *To represent a right-angled triangle in terms of two roots.*

Let the two roots be A and B. We want to represent a right-angled triangle by means of these two roots. Now Pythagoras tells us that the square of the side subtending the right angle is equal to the sum of the squares of the sides round the right angle. The side subtending the right angle is usually called the Hypotenuse and the sides round the right angle are called the perpendicular and the base. The problem thus requires us to express in terms of the two roots three squares such that one is the sum of the two others, and to identify the side of the largest square with the hypotenuse and the sides of the other two with the perpendicular and the base.

Now it has already been shown that the square on the sum of two sides is equal to the square on their difference plus four times the rectangle contained by the two sides.

Let us represent the third proportional between the two given roots A and B: $\dfrac{B \text{ squared}}{A}$. The sum of the extreme values will give the hypotenuse $A + \dfrac{B \text{ squared}}{A}$. Their difference, the base, will be $A = \dfrac{B \text{ squared}}{A}$. [Fig. 4.6].*

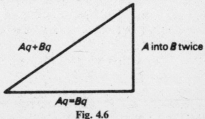

Fig. 4.6

The perpendicular will be B doubled, since the square of B is clearly equal to the product of the extreme values. Multiplying the whole by A, in such a way that by applying it to any side [division] we obtain a quantity of the same type [a length],

the hypotenuse will be A squared $+$ B squared

the perpendicular A into B twice

the base A squared $=$ B squared.†

---

* We must remember that in Vieta's work $=$ indicates the absolute value of a difference. There is no equals sign.

† In the figure, the square of A is written Aq, q being an abbreviation for 'quadratus', meaning squared. The abbreviation c is used for a cube.

Thus a right-angled triangle has been represented in terms of two sides: the hypotenuse is similar to the sum of the squares, the base to their difference and the perpendicular to twice the rectangle.

In the same way one can represent a right-angled triangle in terms of three proportionals. The hypotenuse will be similar to the sum of the extreme values, the base to their difference, the perpendicular to twice the mean.

Deduction: In right-angled triangles the perpendicular is a mean proportional between the sum and the difference of the base and the hypotenuse.

In his edition of Diophantus, published in 1621, Bachet de Méziriac adds the following problem at the end of his Latin commentary:

PROBLEM XX. Invenire triangulum rectangulum, cujus area sit datus numerus. Oportet autem ut quadratus areae duplicatae additus alicui quadratoquadrato, faciat quadratum.*

Bachet's solution is uninteresting, but in the margin of his copy of Bachet's book Fermat wrote some very interesting comments on this proposition.

The numerical value of the area of a right-angled triangle cannot be a perfect square.†

I shall give a proof of this theorem I have discovered; it cost me much toil to work out this proof but proofs of this type will greatly advance our understanding of numbers.

Opposite our translation we give explanations in modern notation.

If the area of a triangle were a square there would be two *bi-quadratics* whose difference would be a square;

The sides of the triangle are $k(m^2 + n^2)$, $k(m^2 - n^2)$ and $2kmn$, where m and n are prime to one another and one of them is even while the other is odd. The area ($\frac{1}{2}bc$) is $k^2mn(m^2 - n^2)$, therefore $mn(m^2 - n^2)$ is a square. Since the three factors are prime to one

---

* To find a right-angled triangle whose area is a given number. As an additional condition we require that the square of twice the area plus the fourth power of some number shall be equal to a square.

† This proposition has been stated in a slightly different form by Leonardo of Pisa, in 1225, in his *Liber Quadratorum*. His proof was, however, faulty.

it follows that there would also be two squares whose sum and difference were also squares.

another each of them must be a square:

$$\text{let } m = p^2, n = q^2,$$

and

$$m^2 - n^2 = p^4 - q^4 = r^2.$$

$$p^4 q^4 = (p^2 + q^2)(p^2 - q^2) = r^2.$$

Since p and q are one even and one odd, like m and n, and are prime to one another, $p^2 + q^2$ and $p^2 - q^2$ are both odd and are prime to one another. Since their product is a square each of these numbers is therefore a square.

Therefore, we should have a square which was the sum of a square and of twice a square, where the sum of the two squares which composed it was also a square.

$$p^2 + q^2 = x^2$$
$$p^2 - q^2 = y^2$$
$$p^2 = q^2 + y^2$$
$$x^2 = 2q^2 + y^2$$

x^2, a square, is the sum of a square, y^2, and twice a square, $2q^2$.

But if a square is the sum of a square and of twice a square its root is also the sum of a square and of twice a square, as I can easily prove.

Also, $y^2 + q^2 = p^2$

$$x^2 = 2q^2 + y^2$$
$$x^2 - y^2 = (x - y)(x + y) = 2q^2$$

and x and y are prime to one another and both odd. $x - y$ and $x + y$ therefore only have the one common factor 2. Therefore

$$x - y = 4u^2 \quad \text{or} \quad 2u^2$$
$$x + y = 2v^2 \quad \text{or} \quad 4v^2.$$

From this we conclude that this root is the sum of the two sides which enclose the right angle of the triangle, one of the squares forming the base of the triangle and twice the other forming its height.

So

$$x = v^2 + 2u^2 \quad \text{or} \quad u^2 + 2v^2$$
$$y = v^2 - 2u^2 \quad \text{or} \quad 2v^2 - u^2 \,\,*$$
$$2q^2 = 8u^2v^2$$

and

$$x^2 = 2q^2 + y^2$$
$$= 8u^2v^2 + (v^2 - 2u^2)^2$$
$$= v^4 + 4u^4$$

---

* In this second case the calculation proceeds exactly as in the first one.

v^2 and $2u^2$ are the sides of a right-angled triangle with sides equal to integral numbers of units, i.e. a Pythagorean triangle.

This right-angled triangle will thus be composed of two square numbers whose sum and difference are also squares.

Like all Pythagorean triangles the one we have just constructed can be produced from two different numbers. That is, since v^2 and $2u^2$ are prime to one another and $2u^2$ is even, there exist two numbers k and l, prime to one another, and one even and the other odd, such that

$$2u^2 = 2kl$$
$$v^2 = k^2 - l^2$$

or

$$kl = u^2$$
$$(k - l)(k + l) = v^2.$$

From the first equation it is clear that k and l are squares, and from the second that their sum and their difference are also squares.

But we shall show that the sum of these two squares is smaller than the sum of the first two whose sum and difference we also supposed to be squares.

These first numbers are p^2 and q^2. Their sum is x^2 and since $x^2 = v^4 + 4u^4$, we have $x^2 > v^2 > k + l$.

Therefore, if we are given two squares whose sum and difference are squares this is equivalent to giving, in integers, two squares which also have the above property, and whose sum is less that that of the first two.

By the same reasoning as before we can now find another sum smaller than that found from the first sum, and by continuing this process indefinitely we shall find smaller and smaller integers which satisfy the same conditions. But this is impossible, since there cannot be an infinite number of integers smaller than some given integer.

This margin is too narrow for the complete proof and all its consequences.

By the same method I have discovered that there is no *biquadratic* triangular number other than unity.

5. Trigonometry

If we want to bore a tunnel for laying mines, to carry out an orderly bombardment, to calculate the parts of a regular fortification, to mark it out on the ground, to build a camp, to draw a map or the plan of a trench, or to decide how to point our artillery, we must necessarily have recourse to trigonometry.
Belidor, 1725

'Trigonometry', wrote Cagnoli, 'is a word taken from the Greek, and it means the measurement of triangles.' The word is, indeed, taken from the Greek. It is not however a Greek word, since although this branch of mathematics is undoubtedly of Greek origin it was not known by its present name until the beginning of the seventeenth century AD. The name, it seems, was invented by the German astronomer Pitiscus, in 1599, when he entitled one of his works *Trigonometria libri quinque*.

The subject itself had probably been studied, in various contexts, from about the third century BC onwards.

Autolycos of Pitane (fl. 310 BC) wrote two short treatises which still survive: one is called *On the Moving Sphere* and the other *On Risings and Settings* [of stars]; while among the surviving works of Euclid there is a treatise on elementary astronomy, called *Phaenomena*.

These three works lead us to suppose that there must have been a treatise on the fixed sphere, but the only Greek works on the subject that have survived date from a much later period.

Theodosius's *Three books on the sphere* date from about 200 BC.

Theodosius defines the sphere as a solid body contained by a surface which is everywhere equidistant from the same internal point. He then shows that any plane section of a sphere is a circle, and that the straight line joining the centre of the sphere to the centre of the section is perpendicular to the plane of the section. He deduces the inequalities which relate the distances from the centre to the sizes of the sections, introduces the idea of a pole, and discusses the properties of a plane which is tangent to the sphere at a given point, as well as the properties of the "greatest circles". The whole of his first book is thus taken from Book III of Euclid's *Elements*.

The second book begins by defining circles tangent to one another on the sphere, and goes on to discuss the properties of parallel, tangent and secant circles etc.

The third book gives a geometrical treatment of problems that are of interest in astronomy.

Menelaus's *Three books on the sphere* are much more sophisticated than Theodosius's treatise. They contain the famous theorem now known as Menelaus's Theorem, which we may state in modern terms as follows:

If the sides of a plane or spherical triangle ABC are cut by a straight line or a great circle in the points L, M and N, then, in the plane case, we have

$$\frac{LA}{LB} = \frac{NA}{NC} \times \frac{MC}{MB}$$

and in the spherical case

$$\frac{\sin LA}{\sin LB} = \frac{\sin NA}{\sin NC} \times \frac{\sin MC}{\sin MB}.$$

This theorem was later to prove very important for spherical trigonometry.

Menelaus also wrote a work in six books about chords in circles. This work, which is now lost, must have been somewhat similar to the passage from Ptolemy quoted below. There may also have been earlier models for this work, dating back at least to the astronomer Hipparchus, in the second century BC.

The best surviving example of Greek trigonometry is to be found in Chapters IX and XI of the first book of the *Almagest* (written by Claudius Ptolemy).

Chapter IX deals with calculating the lengths of chords inscribed in a circle. Chapter XI entitled *Preliminaries to proofs on the sphere* is essentially concerned with Menelaus's Theorem, first in the plane, and then on the sphere:

If, given two straight lines *AB* and *AG*, we draw two more straight lines *BE* and *GD* to meet them and to intersect one another in the point *Z*, I say that the ratio of *GA* to *AE* is composed of the ratio of *GD* to *ZD* and the ratio of *ZB* to *BE* [Fig. 5.1].

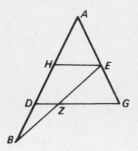

Fig. 5.1

The proposition is proved by drawing through *E* a line *EH*, parallel to *DG*, and then using Thales's Theorem.

The proposition for the sphere is stated:

Let there be drawn on the surface of a sphere arcs of great circles such that the two arcs *BE* and *GD*, drawn to meet the two arcs *AB* and *AG*, intersect one another at the point *Z*, and each arc is less than a semi-

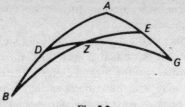

Fig. 5.2

circumference [Fig. 5.2]. I say that the ratio of the chord of double the arc *GE* to the chord of double the arc *EA* is composed of the ratio of the chord of double the arc *GZ* to the chord of double the arc *ZD*, and of the ratio of the chord of double the arc *DB* to the chord of double the arc *BA*.

The proposition is proved by projecting the figure, with the centre of the sphere as vertex, onto the plane *DAG*, and then reducing the proposition to the plane case by means of a lemma (see Fig. 5.3):

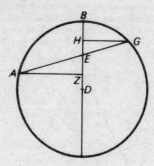

Fig. 5.3

Let *D* be the centre of the circle *ABG* and let us take three points *A*, *B* and *G* on its circumference such that each of the arcs *AB* and *BG* is less than a semi-circle.... Join *AG* and *DEB*. Now the ratio of the straight line subtending twice the arc *AB* to the straight line subtending twice the arc *BC* is the same as the ratio of the straight line *AE* to the straight line *EG*. For let us draw through the points *A* and *G* lines *AZ* and *GH* perpendicular to the line *DB*. Since *AZ* is parallel to *GH*, and the straight line *AEG* is a transversal to these parallel lines, we have $AE:EG = AZ:GH$. But the ratio of *AZ* to *GH* is the same as that of the straight line subtending twice the arc *AB* to the straight line subtending twice the arc *BG*. For each of the perpendiculars is half of the corresponding straight line subtending the arc.

All Ptolemy's astronomical calculations are made, not, as they would today, by using spherical triangles, but by a systematic use of Menelaus's Theorem.

Chapter IX of Book I of the *Almagest* reads as follows [1]:

On the size of Chords in a Circle

With an eye to immediate use, we shall now make a tabular exposition of the size of these chords by dividing the circumference into 360 parts and setting side by side the chords as the arcs subtended by them increase by a half part. That is, the diameter of the circle will be cut into 120 parts for ease in calculation; [and we shall take the arcs, considering them with respect to the number they contain of the circumference's 360 parts, and compare them with the subtending chords by finding out the number the chords contain of the diameter's 120 parts.] But first we shall show how, with as few theorems as possible and the same ones, we make a methodical and rapid calculation of their sizes so that we may not only have the magnitude of the chords set out without knowing the why and wherefore but so that we may also easily manage a proof by means of a systematic geometric construction. In general we shall use the sexagesimal system because of the difficulty of fractions, and we shall follow out the multiplications and divisions, aiming always at such an approximation as will leave no error worth considering as far as the accuracy of the senses is concerned.

Then first let there be the semicircle *ABG* on the diameter *ADG* and around centre *D*, and let straight line *DB* be erected on *AG* at right angles [Fig. 5.4]. Let *DG* be bisected at *E*, and *EB* be joined; and let *EZ* be laid out equal to *EB*, and let *ZB* be joined.

I say that the straight line *ZD* is the side of a regular inscribed decagon, and *BZ* that of a pentagon.

For since the straight line *DG* is bisected at *E* and a straight line *DZ* is added to it,

$$\text{rect. } GZ \cdot ZD + \text{sq. } ED = \text{sq. } EZ \text{ [Eucl. II, 6]}$$
$$= \text{sq. } BE,$$

since $BE = EZ$.

But

$$\text{sq. } ED + \text{sq. } DB = \text{sq. } BE \text{ [Eucl. I, 47]}$$

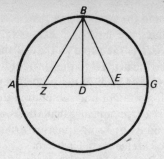

Fig. 5.4

Therefore

$$\text{rect. } GZ \cdot ZD + \text{sq. } ED = \text{sq. } ED + \text{sq. } DB.$$

And, subtracting the common square on *ED*,

$$\text{rect. } GZ \cdot ZD = \text{sq. } DB$$

$$= \text{sq. } DG.$$

Therefore *GZ* is cut at *D* in extreme and mean ratio [Eucl. VI, def. 3]. Since, then, the side of the hexagon and the side of the decagon which are inscribed in the same circle, when they are in the same straight line, cut that line in extreme and mean ratio [Eucl. XIII, 9], and since the radius *DG* is equal to the side of the hexagon [Eucl. IV, 15 coroll.], therefore *ZD* is equal to the side of the decagon.

And likewise, since the square on the side of the pentagon is equal to the square on the side of the hexagon together with the square on the side of the decagon, all inscribed in the same circle [Excl. XIII, 10], and since in the right triangle *BDZ*

$$\text{sq. } BZ = \text{sq. } DB + \text{sq. } ZD$$

where *DB* is the side of the hexagon and *ZD* the side of the decagon the straight line *BZ* is equal to the side of the pentagon.

Since, then, as I said, we suppose the diameter divided into 120 parts, therefore by what we have just established, being half the circle's radius,

$$ED = 30 \text{ such parts,}$$

and

$$\text{sq. } ED = 900;$$

and

$$\text{rad. } DB = 60 \text{ such parts,}$$

and

$$\text{sq. } DB = 3600;$$

and

$$\text{sq. } BE = \text{sq. } EZ = 4500.$$

Then $EZ \doteq 67^p4'55''$* in length, and by subtraction

$$ZD \doteq 37^p4'55''.$$

Therefore, the side of the decagon, subtending an arc of 36° of the whole circumference's 360, will have $37^p4'55''$ of the diameter's 120^p. Since again

$$ZD = 37^p4'55'',$$

$$\text{sq. } ZD = 1375^p4'14'',$$

$$\text{sq. } DB = 3600^p,$$

and sq. ZD + sq. DB = sq. BZ. Therefore, in length,

$$BZ \doteq 70^p32'3''.$$

And therefore the side of the pentagon, subtending an arc of 72° is $70^p32'3''$.

It is immediately clear that the side of the hexagon, subtending an arc of 60° and being equal to the radius, is itself 60^p. And likewise, since the side of the inscribed square, subtending an arc of 90°, is, when squared, double the square on the radius, and since the side of the inscribed equilateral triangle is, when squared, triple the square on the radius, and since the square on the radius is 3600^p, the square on the side of the square will add up to 7200^p, and the square on the side of the equilateral triangle to $10\,800^p$. And so in length

$$\text{chord of arc } 90° \doteq 84^p51'\,10'',$$

and

$$\text{chord of arc } 120° \doteq 103^p55'\,23''.$$

And so these chords are easily got by themselves. Thence it is evident that, with these chords given, it will be easy to get the chords which subtend the supplements, since the squares on them added together are equal to the square on the diameter. For example, since it was shown

$$\text{chord of arc } 36° = 37^p4'\,55'',$$

$$\text{sq. chord of arc } 36° = 1375^p4'\,55'',$$

_____

* From now on, we shall indicate 'parts such as the diameter's 120' by a p-superscript. Thus $67^p4'\,55''$ means '$67 + \frac{4}{60} + \frac{55}{3600}$ parts such as the diameter's 120'. And we shall indicate 'parts such as the circumference's 360' by the ordinary notation for angular degrees. For the measures of arcs and angles exactly correspond. Thus $47°42'40''$ means '$47 + \frac{42}{60} + \frac{40}{3600}$ parts such as the circumference's 360' or 'parts such as 4 right angles' 360'.

and

$$\text{sq. diameter} = 14\,400^P,$$

therefore, for the supplement,

$$\text{sq. chord of arc } 144° \doteqdot 13\,024^P 55' 45'',$$

and, in length,

$$\text{chord of arc } 144° = 144^P 7' 37'';$$

and the others in like manner.

And we shall next show, by expounding a lemma very useful for this present business, how the rest of the chords can be derived successively from those we already have.

For let there be a circle with any sort of inscribed quadrilateral *ABGD*, and let *AG* and *BD* be joined [Fig. 5.5].

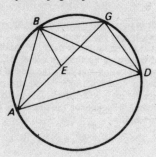

Fig. 5.5

It is to be proved that

$$\text{rect. } AG \cdot BD = \text{rect. } AB \cdot DG + \text{rect. } AD \cdot BG.$$

For let it be laid out such that

$$\text{angle } ABE = \text{angle } DBG.$$

If then we add the common angle *EBD*,

$$\text{angle } ABD = \text{angle } EBG.$$

But also

$$\text{angle } BDA = \text{angle } BGE \text{ [Eucl. III, 21],}$$

for they subtend the same arc. Then triangle *ABD* is equiangular with triangle *BGE*. Hence

$$BG : GE :: BD : AD \text{ [Eucl. VI, 4].}$$

Therefore rect. *BG . AD* = rect. *BD . GE* [Eucl. VI, 16]. Again since angle *ABE* = angle *GBD* and also angle *BAE* = angle *BDG*, therefore triangle *ABE* is equiangular with triangle *BGD*. Hence

$$AB:AE::BD:GD.$$

Therefore rect. *AB . GD* = rect. *BD . AE*. But it was also proved

$$\text{rect. } BG . AD = \text{rect. } BD . GE$$

Therefore also rect. *AG . BD* = rect. *AB . GD* + rect. *BG . AD* [Eucl. II, 1]. Which was to be proved.

Now that this has been expounded, let there be the semicircle *ABGD* on diameter *AD* [Fig. 5.6], and from the point *A* let there be drawn the two

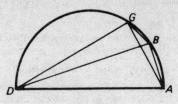

Fig. 5.6

straight lines *AB*, *AG*, and let the length of each of them have been given in terms of such parts as the given diameter's 120; and let *BG* be joined.

I say that *BG* is also given.

For let *BG* and *GD* be joined. Then clearly they are also given because they subtend the supplements. Since, then, the quadrilateral *ABGD* is inscribed in a circle, therefore

rect. *AB . GD* + rect. *AD . BG* = rect. *AG . BD* [proved above].

And rectangle *AG . BD* is given, and also rectangle *AB . GD*. Therefore the remaining rectangle *AD . BG* is also given. And it is now clear to us that, if two arcs are given and the two chords subtending them, than also the chord subtending the difference between the two arcs will be given. And

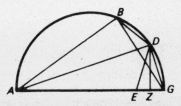

Fig. 5.7

it is evident that by means of this theorem many other chords in arcs which are the differences between arcs directly given; for instance, the chord subtending an arc of 12°, since we have the chords of 60° and 72°.

Again, given any chord in a circle, let it be proposed to find the chord of half the arc of the given chord.

And let there be the semicircle ABG on diameter AG, and let GB be the given chord [Fig. 5.7]. And let the arc be bisected at D, and let AB, AD, BD, and DG be joined. And let DZ be drawn from D perpendicular to AG.

I say that $GZ = $ half $(AG - AB)$.

For let AE be laid out such that

$$AE = AB,$$

and let DB be joined. Since

$$AB = AE,$$

and AD is common, therefore the two sides AB and AD are equal to the two sides AE and AD respectively. And

angle BAD = angle EAD [Eucl. III, 27];

therefore also base BD = base DE. But chord BD = chord GD, and therefore chord $GD = DE$. Since then, in the isosceles triangle DEG, DZ has been dropped from the vertex perpendicular to the base, therefore

$$EZ = GZ \text{ [Eucl. I, 26]}.$$

But $GE = AC - AB$; therefore $GZ = $ half $(AG - AB)$.

And so, since, given the chord of arc BG, chord AE of its supplement is also given [see above], therefore GZ, which is half the difference between AG and FB, is given too. But when the perpendicular DZ is drawn in right triangle AGD, as a consequence right triangle AGD is equiangular with right triangle DGZ [Eucl. VI, 8] and

$$AG:GD: :GD:GZ.$$

Therefore rect. AG . $GZ = $ sq. GD. But rectangle AG . GZ is given, therefore the square on GD is also given. And so the chord GD of half the arc BG will also be given in length.

And so again, by means of this theorem, most of the other chords will be given as subtending the halves of arcs already found. For instance, from the chord of an arc of 12°, there can be gotten the chord of an arc of 6°, and those subtending arcs of 3°, $1\frac{1}{2}°$ and $\frac{3}{4}°$ respectively. And we shall find from calculation that

$$\text{chord of arc } 1\frac{1}{2}° \doteqdot 1^p34'15'',$$

and

$$\text{chord of arc } \tfrac{3}{4}° \doteqdot 0^p47'18''.$$

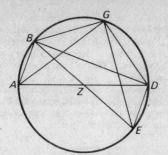

Fig. 5.8

Again, let there be the circle *ABGD* on diameter *AD* with centre at *Z* [Fig. 5.8]. And from *A* let there be cut off consecutively two given arcs, *AB* and *BG*; and let the given chords subtending them, *AB* and *BG*, be joined.

I say that, if we join *AG*, then *AG* will be given also.

For let the circle's diameter *BZE* be drawn through *B*, and let *BD*, *DG*, *GE* and *DE* be joined. Then, from this it is clear that, by means of *BG*, chord *GE* is given; and, by means of *AB*, chords *BD* and *DE* are given [see above]. And by things we have already proved, since *BGDE* is a quadrilateral inscribed in a circle, and *BD* and *GE* are the diagonals, the rectangle contained by the diagonals is equal to the sum of the rectangles contained by opposite sides [see above]. And so, since the rectangles *BG . GE* and *BG . DE* are given, therefore the rectangle *BE . GD* is given also. But the diameter *BE* is given too, and the remaining side *GD* will be given. Therefore the chord *AG* of the supplement will be given also. And so, if two arcs and their chords are given, then by means of this theorem the chord of both these arcs together will be given.

And it is evident that, by continually combining the chord of an arc $1\frac{1}{2}°$ with those so far set out and calculating the sums, we shall inscribe all those chords which, when doubled, are divisible by three; and only those chords will still be skipped which fall within $1\frac{1}{2}°$ intervals.* For there will be two such chords skipped in each interval, since we are carrying out this inscribing of chords by successive additions of $\frac{1}{2}°$. And so if we could compute the chord subtending an arc of $\frac{1}{2}°$, then this chord, by addition to and subtraction from, the chords which are separated by $1\frac{1}{2}°$ intervals and have already been given, will fill in all the rest of the intermediate chords. But since, given any chord such as that subtending an arc of $1\frac{1}{2}°$, the chord of a third of the arc is in no way geometrically given

---

* See below p. 167*ff* in our chapter about the problem of trisecting an angle.

(and if it were possible, we could then compute the chord of an arc of $\frac{1}{2}°$), therefore we shall first look for the chord of an arc of 1° by means of chords subtending arcs of $1\frac{1}{2}°$ and $\frac{3}{4}°$. We shall do this by presenting a little lemma which, even if it may not suffice for determining their sizes in general, can yet in the case of the very small chords keep them indistinguishable from chords already determined.

For I say that, if two unequal chords are inscribed in a circle, the greater has to the less a ratio less than the arc on the greater has to the arc of the less.

For let there be a circle *ABGD* [Fig. 5.9]; and let unequal chords be inscribed in it, *AB* the less and *BG* the greater.

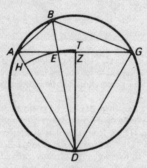

Fig. 5.9

I say that

chord *BG*:chord *AB* < arc *BG*:arc *AB*.*

For let angle *ABG* be bisected by *BD*, and let *AEG*, *AD*, and *GD* be joined. And since angle *ABG* has been bisected by the straight line *DEB*,

chord *GD* = chord *AD* [Eucl. III, 26, 29],

and *GE* > *AE* [Eucl. VI, 3].

Then let *DZ* be dropped from *D* perpendicular to *AEG*. Now since

$$AD > DE,$$

$$DE > DZ,$$

therefore the circle described with centre *D* and radius *DE* cuts *AD* and falls beyond *DZ*. Then let the circle *HET* be drawn and the straight line *DZT* be produced. And since

sector *DET* > triangle *DEZ*

---

* A proposition used earlier by Aristarchus of Samos, Archimedes and, in part, by Euclid.

and

triangle *DEA* > sector *DEH*.

therefore triangle *DEZ*:triangle *DEA* < sector *DET*:sector *DEH* [Eucl.
V, 8]. But triangle *DEZ*:triangle *DEA*: :*EZ*:*AE* [Eucl. VI, 1], and sector
DET:sector *DEH*: :angle *ZDE*:angle *EDA*. Therefore *EZ*:*AE* < angle
ZDE:angle *EDA*. Then *componendo AZ*:*AE* < angle *ZDA*:angle *EDA*.
And doubling the antecedents

GA:*AE* < angle *GDA*:angle *EDA*.

And *separando GE*:*AE* < angle *GDB*:angle *BDA*. But *GE*:*AE*: :*BG*:*AG*
[Eucl. VI, 3], and angle *GDB*:angle *BDA*: :arc *BG*:arc *AB* [Eucl. VI, 33].
Therefore chord *BG*:chord *AB* < arc *BG*:arc *AB*.

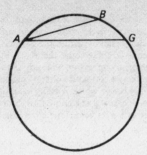

Fig. 5.10

Now, then, with this laid down, let there be the circle *ABG* [Fig. 5.10],
and let the two chords *AB*, *AG* be inscribed in it. And first let *AB* be given
as subtending an arc of $\frac{3}{4}°$, and *AG* an arc of 1°.

Since chord *AG*:chord *AB* < arc *AG*:arc *AB*, and arc *AG* = $1\frac{1}{3}$ (arc *AB*),
therefore chord *AG* < $1\frac{1}{3}$ (chord *AB*). But it was proved [see above]

chord *AB* = 0^{p}47′8″.

Therefore chord *AG* < 1^{p}2′50″, for 1^{p}2′50″ ≑ $1\frac{1}{3}$(0^{p}47′8″).

Again, with the same figure, let chord *AB* be given as subtending an arc
of 1°, and chord *AG* an arc of $1\frac{1}{2}°$.

Likewise then, since arc *AG* = $1\frac{1}{2}$ (arc *AB*),

chord *AG* < $1\frac{1}{2}$ (chord *AB*).

But we proved [see above] chord *AG* = 1^{p}34′15″. Therefore chord
AB > 1^{p}2′50″, for 1^{p}34′15″ ≑ $1\frac{1}{2}$(1^{p}2′50″).

And so, since it has been proved that the chord of an arc of 1° is both
greater and less than the same number of parts, clearly we shall have chord
of arc 1° = 1^{p}2′50″;

and by means of earlier proofs we saw

$$\text{chord of arc } \tfrac{1}{2}^\circ \fallingdotseq 0^p 31' 25''.$$

And the remaining intervals will be filled in as we have just said.* For example, in the first interval we find the chord subtending an arc of 2° by adding $\tfrac{1}{2}^\circ$ and $1\tfrac{1}{2}^\circ$, and the chord subtending an arc of $2\tfrac{1}{2}^\circ$ by subtracting $\tfrac{1}{2}^\circ$ from 3°, and so on for the rest.

So the business of chords in a circle can be easily handled in this way, I think, And as I have said, in order to have the magnitudes set out immediately to hand, we shall draw up tables of 45 rows each, for symmetry's sake. And the first column will contain the magnitudes of the arcs increasing by $\tfrac{1}{2}^\circ$, and the second column will contain the magnitudes of the chords subtending them in terms of the diameter's assumed 120 parts. The third column will contain the thirtieth of the increase of the chords as the corresponding arc increases by $\tfrac{1}{2}^\circ$, so that we may have a mean addition, accurate for the senses, for each increase of $\tfrac{1}{60}^\circ$ in the corresponding arcs, and so be able to calculate readily the chords falling within the $\tfrac{1}{2}^\circ$ intervals. And it is to be remarked that by means of these same theorems, if we should suspect some typographical error in connection with any of the chords computed here, we can easily check and correct it either by means of the chord of an arc double the arc of the chord which is being examined, or by means of the difference of certain other given chords, or by means of the chord subtending the supplement. And here is the table [Fig. 5.11]:†

Table of Chords Inscribed in a Circle

Arcs		Chords			Thirtieths of the differences			
degrees	minutes	parts	first	second	parts	first	second	third
0	30	0	31	25	0	1	2	50
1	0	1	2	50	0	1	2	50
1	30	1	34	15	0	1	2	50
2	0	2	5	40	0	1	2	50
2	30	2	37	4	0	1	2	48
3	0	3	8	28	0	1	2	48

Fig. 5.11

---

* The chord $30 = 2 \sin 15$; $0^p 31' 25'' = 0.0087268\ldots$ of the radius. Modern tables give the value $0.0087266\ldots$

† We reproduce the beginning of this table, in translation and in its original form, as an example of Greek numerals.

KANONION TΩN EN KYKΛΩ EYΘEIΩN

ΠΕΡΙΦΕΡΕΙΩΝ		ΕΥΘΕΙΩΝ			ΕΞΗΚΟΣΤΟΝ			
Μοιρῶν		M	Π	Δ	M	Π	Δ	T
δ	λ	δ	λα	κε	ō	α	β	ν
α	δ	α	β	ν	ō	α	β	ν
α	λ	α	λδ	ιε	ō	α	β	ν
β	ō	β	ε	μ	ō	α	β	ν
β	λ	β	λζ	δ	ō	α	β	μη
γ	ō	γ	η	κη	ō	α	β	μη

Fig. 5.11 *continued*

Ptolemy frequently used the chord of twice the arc
considered, particularly in connection with Menelaus's
Theorem. In about the fourth century AD the Hindus
systematically employed half the chord of the doubled arc,
and they gave this quantity a name we translate as 'sine'.
Aryabhata describes a curious method of drawing up a table
of sines. It gives a rather crude table, with the angles
increasing in steps of $3\frac{3}{4}°$.

Many Arab mathematicians computed trigonometric
tables; among others al-Khovarizmi and al-Battani and,
in the second half of the ninth century AD, Aboul-Wefa,
who described a new method of computing a table of
sines.

Aboul-Wefa made important contributions to the progress of trigonometry. For instance, the sexagesimal value
he gives for sin 30' is correct to eight places of decimals.
Not only did he compute sines but he also introduced
tangents and secants (the reciprocals of cosines). He used
computational rules or canons equivalent to the following
formulae.

$$\frac{1}{2}\frac{\text{versin } u^*}{\sin u/2}$$

$$\frac{\text{chord } u}{\text{chord } u/2}$$

$$= \frac{\sin u/2}{R}$$

$$= \frac{\text{chord }(\pi R - u/2)}{R}$$

$$\frac{2R - \text{chord }(\pi R - u)}{\text{chord } u/2}$$

$$\frac{\sin u}{\sin u/2}$$

$$= \frac{\text{chord } u/2}{R}$$

$$= \frac{\sin (\pi/2)R - u/2}{\frac{1}{2}R}$$

From geometrical considerations Aboul-Wefa obtains

$$\sin(a \pm b) = \sqrt{\sin^2 a - \frac{\sin^2 a \sin^2 b}{R^2}}$$
$$\pm \sqrt{\sin^2 b - \frac{\sin^2 a \sin^2 b}{R^2}},$$

but he then goes on to derive a more elegant formula, which we should write in modern notation as

$$\sin (a \pm b) = \frac{\sin a \cos b \pm \cos a \sin b}{R}.$$

In computing his tables Aboul-Wefa uses the fact that in the first quadrant, if b is positive,

$$\sin (a + b) - \sin a < \sin a - \sin (a - b).$$

This enables him to assess the accuracy of his results.

Al-Battani, who died in AD 929, was acquainted with a formula equivalent to one of the fundamental formulae of modern spherical trigonometry, namely the formula

$$\cos a = \cos b \cos c + \sin b \sin c \cos A,$$

in which a, b and c are the sides of a spherical triangle and A is the angle opposite the side a.

---

* Versin $u = R - \cos u$. The radius, which is not taken as unity, appears in all the formulae. It is numerically equal to 60. The sines, tangents etc. are lengths of lines not ratios.

PLATE VIII. The use of a geometric square. From S. Munster, *Rudimenta* (1557).

B.N.

Ibn-Yunus, who died in Egypt in the year AD 1009, derived a formula analogous to what we should write as:

$$\cos a \cos b = \tfrac{1}{2}[\cos(a + b) + \cos(a - b)]$$

The *Treatise on the Quadrilateral* of Nasir al-din al-Tusi provides an admirable example of how the Arabs of the Middle East used tables in their work in plane trigonometry and, more particularly, in spherical trigonometry. Al-Tusi shows how to solve plane triangles by decomposing them into right-angled triangles, then proves that the sides of a triangle are proportional to the sines of the opposite angles, and goes on to deduce all the formulae we should use today in solving right-angled spherical triangles. General spherical triangles are solved either by decomposing them into right-angled triangles or by means of what is called a supplementary figure. Al-Tusi also shows that the sines of the sides are proportional to the sines of the opposite angles (a result known to Aboul-Wefa). When the triangle is defined by its three angles it is solved by means of a supplementary triangle. In the West this idea was redis-covered by Vieta. It is equivalent to using a supplementary trihedral angle.

In the West, interest in trigonometrical work did not really begin to revive until the fourteenth century, but the practical geometry of the Middle Ages does deserve to be

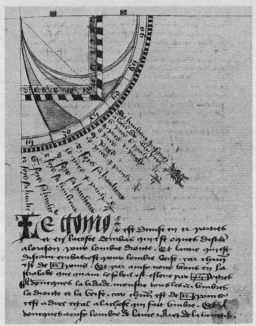

PLATE IX. An astrolabe, from an anonymous manuscript in the Bibliothèque Nationale.

mentioned briefly. We have had occasion earlier to refer to the geometric square. Plate VIII shows how this instrument was used. The engraving, taken from the *Rudimenta* of Sebastian Munster, shows an observer, who requires to measure the width of a river and has climbed a ladder. The ray from his eye passes the point marked 6 on the lines of the *umbrae versae* on the geometric square and he deduces from this that the width of the river is twice the height of the ladder.

Over the centuries instruments improved, becoming both more specialized and more numerous. In 1702 Manesson

Mallet described trigonometry as: "the measurement of distances by means of Geometrical Instruments such as Pegs, Cords, the Graphometer, the Geometric Square, the Sector, the Astrolabe, the Compass, Jacob's Staff, the Plane Table and also by Sines and Logarithms".

We have already described some of the instruments and methods to which Manesson Mallet refers.

Plate IX, taken from an anonymous manuscript in the Bibliothèque Nationale in Paris, shows an astrolabe.

The instrument consists of a quadrant (a quarter of a circle) divided into ninety degrees, and it is interesting to notice that the shapes of the numbers are different from modern ones. Inside the quadrant there are lines for the hours, each of them an arc of a circle. There is also a geometric square, and on the upper edge of the instrument there are two pin-holes for taking sightings. The accompanying text tells us: "The gnomon is divided into twelve parts along the lower side, which is parallel to the horizon and gives the *umbra recta*. And the other, vertical, side is for the *umbra versa*. This is similarly marked with twelve units. And thus we can see by the astrolabe that when the sun is at a height of 45 degrees then the alidade [set of sights] shows the two *umbrae*, the *umbra recta* and the *umbra versa*, equally as 12 units, that is, equal to the object casting the shadow. So then the shadow of a spear is equal to the spear."

Below the rim, working out from the middle towards the left, we read:

"The height to the eye is	La hauteur jusques à
6 feet	l'oeil est de 6 pieds
6 feet, 6 inches 6/11	6 pieds, 6 pouces 6/11
7 feet, 2 inches 2/5	7 pieds, 2 pouces 2/5
8 feet	8 pieds
9 feet	9 pieds

10 feet, 3 inches 3/7	10 pieds, 3 pouces 3/7
2 times its height	2 fois sa hanteur
14 feet, 4 inches 4/5	14 pieds, 4 pouces 4/5
3 times its height	3 fois sa hauteur
4 times its height	4 fois sa hauteur
6 times its height	6 fois sa hauteur
12 times its height	12 fois sa hauteur "

We can also see the set of sights positioned along the line bisecting the instrument. The alidade could be turned about the centre of the quadrant to allow sightings to be taken obliquely.

This very simple and portable instrument had many uses, such as measuring distances, heights of towers, depths of wells, and solving other such problems of practical geometry. It could also be used to measure the altitude of the sun or of the stars so as to determine the time either by day or by night.

Today there are specialists to deal with many different problems which those who lived in the Middle Ages would have had to solve for themselves. They could not avail themselves of the Speaking Clock, or even necessarily of the ordinary clock in the church tower, nor could they all carry a chronometer or a Nuremberg egg,* so they required some means of determining the time for themselves, and since they would have used not proper maps but merely approximate itineraries they would also have required some means of determining their geographical position.

The crude instruments we have described required considerable skill on the part of the user, and if a 'clerk' knew only a little about astronomy, arithmetic and geometry he at least knew that little very thoroughly.

---

* A type of pocket clock made in Nuremberg by Peter Hole as early as 1490. So called because of its oval shape. J.V.F.

We should note that the manuscript uses the words *gnomon*, *horizon* and *astrolabe*, which are Greek, and the word *alidade* which is Arabic. This is a further indication of the origins of the writer's knowledge of astronomy and geometry.

The *umbra recta* and *umbra versa* developed into the modern *tangent* and *cotangent*.

From the Geometric Square and the Quadrant, however, there is still a long way to go before we come to real trigonometry. The primitive instruments we have described merely bear witness to the everyday activities of ordinary people.

However, the science of astronomy, or rather astrology, was cultivated on a much higher level. We have already had occasion to mention the high esteem in which astrology, particularly judicial astrology, was held in the Middle Ages. Great men had horoscopes drawn up for them, and despite the protests of Nicole Oresme Charles V (the Wise) took counsel from his physician Maître Chrétien Gervais, an adept in astrology. The King, indeed, helped him to found his famous Collège de Maître Gervais which for some time was the only college to have a chair of mathematics.

The undisputed authority on astrology was Ptolemy. His work, even more than Euclid's *Elements*, was the subject of a multitude of commentaries and the material for numerous abridged editions. It was Ptolemy's astrology which excited this interest, but in order to understand the astrology scholars were obliged to study his astronomical work as well, and the science of astronomy therefore benefited from the prevailing interest in astrology.

Astrology continued to be important even as late as the seventeenth century. Morin (1583–1656) exercised great influence in France and men such as Kepler (1571–1630) and Galileo (1564–1642) cast horoscopes, the latter even drew up Petrarch's horoscope, at the request of Jean Beaugrand.

It was from such dubious origins as these that Western astronomy and trigonometry originated in the fourteenth century. The most creative scholar of the following century was Regiomontanus (Johannes Müller, 1436–1476), who was well-versed in Greek and Arab astronomy. He made use of tangents in his work.

Regiomontanus's short treatise *Compositio Tabularum Sinuum* contains seven propositions which we may summarize in modern notation as follows:

1. To show that $\cos A = \sqrt{R^2 - \sin^2 A}$ [R is the radius of the circle]. Here sines and cosines are not ratios but lengths;

2. To calculate sines from 15′ in steps of 15′;

3. To show that

$$\frac{2 \sin^2(A/2)}{R} = \text{versin } A = R - \cos A;$$

4. To find the length of the side of a regular decagon or a regular pentagon inscribed in a circle;

5. To show that

$$\text{chord } (A - B) = \sqrt{(\sin A - \sin B)^2 + (\cos A - \cos B)^2}.$$

Regiomontanus derives this formula in order to find the side of the regular pentadecagon (a fifteen-sided figure), but he notes that the formula is in fact a general one;

6. To show if $\pi/2 > A > B$ then $A:B > \sin A:\sin B$.

This is analogous to Ptolemy's proposition on chords, but Regiomontanus notes that his own formula is a more convenient one to use interpolation. Once sines have been calculated exactly for arcs in steps of 15′ we may suppose that the increase in the sine is proportional to that in the arc and can thus use a simple rule of three (i.e. linear interpolation). Regiomontanus notes that almost everywhere that he has compared his new method with that of Ptolemy the results have been in perfect agreement.

The seventh proposition, which is to find the sine of an arc from tables, is purely practical. There follow two tables, one in which the radius is 60 000 and another in which it is 100 000.

Regiomontanus's most important work, his *De Triangulis libri quinque*, which was written in 1464 and printed in 1533, is a compendium of the trigonometry of the time. It contains a table in which the radius is 10 000 000.

Vieta's earliest mathematical work was concerned with trigonometry, and he is in fact the first to treat the subject in a modern manner. Unfortunately, the work lies beyond the scope of this book.

Confining ourselves to more elementary matters, we shall now turn to Renaissance methods of solving plane triangles.

The basic formula used is the Sine Rule, which states that the sides of the triangle are proportional to the sines of the opposite angles. The most elegant method of deriving this formula is that of Vieta, who uses the circumscribed circle.

Right-angled triangles are solved in the same way as they would be solved today.

Scalene triangles are broken down into right-angled triangles. The formula

$$(a + b):(a - b) = tg\frac{A + B}{2}:tg\frac{A - B}{2}$$

is also employed.

This procedure seems to be derived from Thomas Fink's *Geometrica Rotundi* of 1591. It was Fink who first used the names tangent and secant.

The Arabs had called tangents "shadows" (translated as *umbrae*), and in the West they had been called *Fecond* and then *Prosinus*. Secants were called *Hypotenuse of the fecond* and then *transinus*.

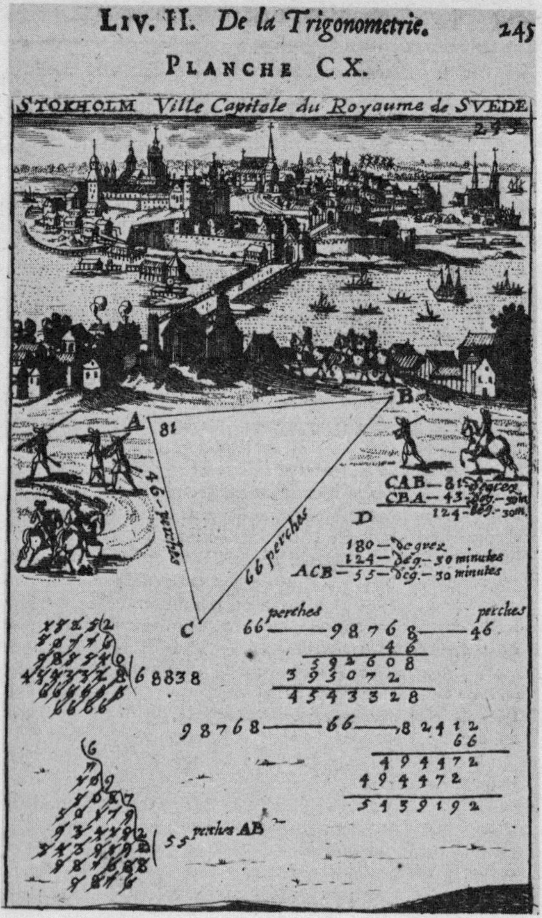

B.N.

PLATE X. From Manesson Mallet's *Géométrie Practique* (1702).

Plate X shows a problem from Manesson Mallet's *Practical Geometry* (1702). It illustrates how a simple triangle was solved:

"Suppose we wish to calculate the distance between *A* and *B*, the side *AC* being assumed to be 46 perches long, the side *CB* 66 perches and the angle *CAB* being 81 degrees".

This is the 'doubtful case' of a modern textbook. It merely requires several applications of the rule of three. All the calculations are shown in the engraving.

Manesson Mallet uses a table of trigonometrical lines in which the radius, 'the total sine', is 100 000.

The sine of 81 degrees is therefore 98 768. An application of the rule of three gives the sine of *B*:

$$98\ 768 \times 46/66 = 68\ 838$$

which gives $B = 43°30'$.

Subtracting from 180° we then obtain $C = 55°30'$, whose sine is 82 412.

Applying the rule of three once more, we obtain *AB*:

$$AB = 82\ 412 \times 66/98\ 768 = 55\ \text{perches}.$$

6. Duplication of the Cube and Trisection of the Angle

> *Nous avons par le mesme moyen la solution de tous les problèmes qui de leur nature sont solides, lesquels en l'Analyse spécieuse, par des préparations convenables, se réduisent a l'un de ces deux esgalitez;*
>
> *A cube esgal a B solide,*
>
> *et B plan par A, moins A cube esgal a Z solide dont nous pourrons quelque jour traiter plus amplement.**

6.1. Duplication of the cube or the Delian problem

We shall begin with a letter from Eratosthenes to King Ptolemy III. (The letter has been shown not to be authentic but it certainly dates from Ancient times since it is quoted by Eutocius) [1].

To King Ptolemy Eratosthenes sends greeting.

They say that one of the ancient tragic poets represented Minos as preparing a tomb for Glaucus, and as declaring, when he learnt it was a hundred feet each way: 'Small indeed is the tomb thou hast chosen for a royal burial. Let it be double, and thou shalt not miss that fair form if thou

---

* In the same way we can solve all problems which are essentially concerned with solids, which by the use of symbols can be reduced to the form of one of these two equations:

A cubed is equal to *B* solid,

and *B* plane times *A*, minus *A* cubed, is equal to *Z* solid, an equation we shall deal with more fully later. Roberval, in Mersenne's *Harmony of the Universe* (Harmonie Universelle), 1637.

quickly doublest each side of the tomb.' He seems to have made a mistake, for when the sides are doubled, the surface becomes four times as great and the solid eight times. It became a subject of inquiry among geometers in what manner one might double the given solid, while it remained the same shape, and this problem was called the duplication of the cube; for, given a cube, they sought to double it. When all were for a long time at a loss, Hippocrates of Chios first conceived that, if two mean proportionals could be found in continued proportion between two straight lines, of which the greater was double the lesser, the cube would be doubled, so that the puzzle was by him turned into no less a puzzle. After a time, it is related, certain Delians, when attempting to double a certain altar in accordance with an oracle, fell into the same quandary, and sent over to ask the geometers who were with Plato in the Academy to find what they sought. When these men applied themselves diligently and sought to find two mean proportionals between two given straight lines, Archytas of Taras is said to have found them by the half-cylinders, and Eudoxus by the so-called curved lines; but it turned out that all their solutions were theoretical, and they could not give a practical construction and turn it to use, except to a certain small extent Menaechmus, and that with difficulty. An easy mechanical solution was, however, found by me, and by means of it I will find, not only two means to the given straight lines, but as many as may be enjoined.

By this means we can convert any solid enclosed by parallelograms into a cube, or convert a solid of one form into another, or increase the volume of a solid while keeping it the same shape. We can do the same for altars and for temples. And we can transform into cubes the measures we use for liquids or for dry materials, such as the metretes and the medimnos,* and from the sides of the cubes we shall know how big these measures are.

My results will also be of use to any who wish to increase the size of catapults and ballistas; for everything must be increased in proportion, thickness, height and the holes and the ropes which pass through them, if the range of the catapult is to be increased proportionately, and this is impossible without a finding means. . . .

We must not regard this letter as completely reliable evidence but it does state the problem fairly clearly.

Duplication of the cube is equivalent to finding an exact value of $\sqrt[3]{2}$. Solving the more general problem, involving any given ratio, is equivalent to solving the equation $x^3 = a$. The problem of finding two mean proportionals between two given lines a and b is that of finding two

---

* metretes ($\mu\varepsilon\tau\rho\eta\tau\dot{\eta}\varsigma$): measure of about 40 litres. medimnos ($\mu\varepsilon\delta\iota\mu\nu o\varsigma$): measure of about 50 litres.

lines x and y such that

$$\frac{x}{a} = \frac{y}{x} = \frac{b}{y} \quad \text{i.e.} \quad \left(\frac{x}{a}\right)^3 = \frac{x}{a} \times \frac{y}{x} \times \frac{b}{y} = \frac{b}{a}.$$

Solving this problem is therefore equivalent to taking the cube root of b/a, and if $b/a = 2$ this latter problem is that of duplicating the cube.

In the letter we have just quoted, the pseudo-Eratosthenes describes Hippocrates of Chios (fl. 450–430 BC) as having discovered that these two problems were identical, but it is likely that this had in fact been known for some time. It might well have been discovered by the Pythagoreans (the first Greek mathematicians to concern themselves with ratios), or even by the Babylonians.

The last part of the letter to Ptolemy describes a practical application, for which an approximate method of finding the cube root would suffice. Heron of Alexandria proceeded as follows (we require a rational approximation to $\sqrt[3]{100}$):

Take the nearest cube numbers to 100 both above and below; these are 125 and 64.

Then $125 - 100 = 25$ and $100 - 64 = 36$.

Multiply 5 into 36; this gives 180. Add 100, making 280. [Divide 180 by 280]; this gives $\frac{9}{14}$. Add this to the side of the smaller cube: this gives $4\frac{9}{14}$. This is nearly as possible the cube root ['cubic side'] of 100 units [2].

The cube root of 100 is 4·6416 ... Heron's method gives $4\frac{9}{14}$, which is 4·6444 ... This is a good approximation, considering how simply it was obtained.

Heron's terse description of his method may, however, cause confusion: when he says "add 100" this 100 is not the given number but the product of the difference between the larger cube and the given number ($125 - 100 = 25$) by the side of the smaller cube (4).

Let us use this method to find the cube root of 2.

$$1^3 = 1 \qquad 2 - 1 = 1$$
$$2^3 = 8 \qquad 8 - 2 = 6$$

$$2 \times 1 = 2 \qquad 1 \times 6 = 6$$
$$2 + 6 = 8$$
$$\tfrac{2}{8} = \tfrac{1}{4}$$

The root is $1\tfrac{1}{4} = 1.25$ (the exact root is $1.2599\ldots$).

This type of interpolation can be justified theoretically.

Suppose we are required to find the cube root of A. Let us call this root a. Then if a lies between two integers p and $q = p + 1$, and we can write $p = a - x$, $q = a + y$, where $x + y = 1$.

Now,

$$p^3 = (a - x)^3 = a^3 - 3a^2x + 3ax^2 - x^3.$$

Writing A for a^3 and neglecting terms in x^3 we obtain

$$A - p^3 = D \simeq 3ax(a - x) = 3axp.$$

In the same way we obtain

$$q^3 = (a + y)^3 = a^3 + 3a^2y + 3ay^2 + y^3.$$

Neglecting y^3 we have

$$q^3 - A = E \simeq 3ay(a + y) = 3ayq.$$

Therefore

$$\frac{x}{y} \simeq \frac{qD}{pE}, \qquad x = \frac{x}{x + y} \simeq \frac{qD}{qD + pE},$$

and

$$a = p + x \simeq p + \frac{qD}{qD + pE}.$$

We do not as yet know exactly how or when this method was invented, but it might be a very old method, perhaps even derived from Babylonian work.

From the time when positional arithmetic was introduced up until the seventeenth century, cube roots, correct to the nearest integer, were found by a method analogous to that used for square roots. The figures were divided up into groups of three, then the first figure of the root was found by using the cubes of the first ten integers, and the following figures were found by dividing by three times the square of the root. This is the method used by Jean Trenchant in his *Arithmetic* (1557) to find the cube root of 99 252 847. Trenchant first divides the figures into groups of three:

$$99/252/847.$$

Now to find the root of the number arranged in this way you must begin at the left with the first group 99, and write down after the bracket the cube root of the largest perfect cube contained in this number, i.e. 4. Then subtract its cube, that is, 64, from 99, and you will be left with 35 under 99, and the remaining part of the cube will be 35 252 847.

$$
\begin{array}{rrr}
99 & 252 & 847(4 \\
64 & \dots & \dots \\
\hline
35 & 252 & 847
\end{array}
$$

This applies to the extraction of any cube root.

Now to find the second figure of the root: square the root you have found, 4, which gives you 16: multiply this 16 by three, which gives you 48: write two noughts after it, or suppose they are there (for the figure 4 of the root is understood to be 40 since we are looking for the next figure) and you will have 4 800: which you will write as divisor under the second group, putting the two noughts under 52 and the figures 48 in order to the left. This shows to begin with how many times 4, or 48, goes into the number above it, which is 6 times: so write 6 for the second figure of the root.

The author goes on to show in detail the following two stages of the calculation:

```
        99│252│847 (46          28800
        64│...│...               4320
        35│252│847                216
divisor  4│800                  33336
        33│336
         1│916│847

        99│252│847 (463        1904400
        64│...│...               12420
        35│252│847                  27
         4│800                 1916847
        33│336
         1│916│847
           634│800
         1│916│847
```

When there is no exact root, various methods are employed to obtain an approximate value.

From the fifth century onwards Hindu mathematicians made use of the approximation $\sqrt[3]{a^3 + b} \simeq a + b/3a^2$, where a^3 is the largest perfect cube less than $a^3 + b$.

This formula is a special case of Newton's method for finding approximate roots of more complicated equations. In some anonymous fifteenth-century manuscripts, and also in the work of Tartaglia, we find the formula

$$\sqrt[3]{a^3 + b} \simeq a + \frac{b}{3a^2 + 3a}.$$

Leonardo of Pisa uses $a + b/(3a^2 + 3a + 1)$, which is what one would obtain from linear interpolation between the numbers shown in a table of cubes.

Jean Trenchant uses this last method to find the cube root of 26, and obtains $2\frac{18}{19}$. He then notes that if the remainder is 1 we can use the approximation $a + b/3a^2 : \sqrt[3]{9} \simeq 2\frac{1}{12}$.

However, Trenchant goes on to describe a method (used earlier by Oronce Finé) which anticipates the use of decimal fractions:

> To any number which does not have an exact root you add as many noughts as you need to make your approximation to the roots as close as you wish, and you continue your method of extraction, taking no account of the remainder, since it will be negligible. . . .
>
> Knowing that 2 579 has not got an exact root I added six noughts: using these I continued my process of extraction and I found 1 371 as remainder for 2 579 000 000. I divided it by 100 and it became $13\frac{71}{100}$, which is the root of 2 579, approximately.

The process of extracting cube roots and, as a special case, doubling the cube, have thus never presented any very great problems as far as practical, approximate computation was concerned; but it was a very different matter in theoretical work.

Today we can easily show that the cube root of a rational number is not, in general, rational.

For let P/Q be a fraction expressed into its lowest terms, i.e. P and Q are assumed prime to one another. Let us suppose that this fraction has a rational cube root, which we can express in its lowest terms as p/q. Now $(p/q)^3 = p^3/q^3$ and since p and q are prime to one another p^3 and q^3 must also be prime to one another. Therefore the fraction p^3/q^3 has been expressed in its lowest terms, and since $p^3/q^3 = P/Q$ we can deduce that $P = p^3$ and $Q = q^3$.

Therefore the only fractions which are perfect cubes are those which, when expressed in their lowest terms, have numerators and denominators which are both perfect cubes.

We should note that Euclid uses exactly this argument in Books VII and VIII of the *Elements*, though he expresses it somewhat differently.

Thus by the third century BC, or possibly by the fourth or even the fifth century, the Greeks had gone some way towards solving the problem of the duplication of the cube, having shown that *the side of the cube and the side of the double cube are not commensurable with one another.*

It was, however, not until the nineteenth century AD that, in the light of Abel's work on algebraic equations, Wantzel was at last able to prove what had long been suspected: that the problem of the duplication of the cube could not be solved by means of a straight edge and compasses alone.

There follows a brief survey of the geometrical methods for duplicating the cube, or constructing the two means, which were proposed between the fourth century BC and the seventeenth century AD.

The various ancient solutions are described by Pappus, in Book III of his *Collection*, and by Eutocius of Askelon in his commentary on Archimedes's treatise *On the Sphere and Cylinder*.

Eutocius ascribes the following solution to Plato [3]:

Given two straight lines, to find two mean proportionals in continuous proportion.

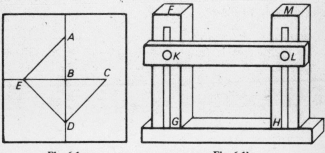

Fig. 6.1a   Fig. 6.1b

Let the two given straight lines be *AB*, *BC*, perpendicular to each other, between which it is required to find two mean proportionals. Let them be produced in a straight line to *D*, [Fig. 6.1a] let the right-angle *FGH* be constructed [Fig. 6.1b], and in one leg, say *FG*, let the ruler *KL* be moved in a kind of groove in *FG*, in such a way that it remains parallel to *GH*. This will come about if another ruler be conceived fixed to *HG*, but parallel to *FG*, such as *MH*. If the upper surfaces of *FG*, *MH* are grooved with axelike grooves, and there are notches on *KL* fitting into the aforementioned grooves, the motion of *KL* will always be parallel to *GH*. When this instrument is constructed, let one leg of the angle, say *GH*, be placed so as to touch *C*, and let the angle and the ruler *KL* be turned about until the point *G* falls upon the straight line *BD*, while the leg *GH* touches *C*, and the ruler *KL* touches the straight line *BE* at *K*, and in the other part touches *A*, so that it comes about, as in the figure, that the right angle takes up the position of the angle *CD*, while the ruler *KL* takes up the position *EA*. When this is done, what was enjoined will be brought about. For since the angles at *D*, *E* are right, $CB:BD = DB:BE = EB:BA$.

Eutocius presents the method somewhat clumsily and makes it seem "mechanical" rather than rigorous. However, if *D* moves uniformly along its half of a straight line, *E* and *A* will always move in the same direction along their lines [the angles at *D* and *E* are assumed to remain right angles] and *A* will pass once and only once through any given point, if we accept the principle of continuity. The method is thus not "mechanical" in the sense of being empirical and approximate and should really be categorised as what Descartes called a "geometrical" method.

The following methods should also be classed with the one ascribed to Plato:

Apollonius's construction

Let the two given straight lines be *AB* and *AC*, which are perpendicular to one another (Fig. 6.2). We construct the

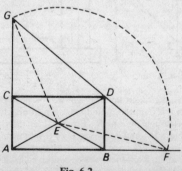

Fig. 6.2

rectangle *ABDC*, with centre *E*. We then draw a circle with centre *E*, to cut *AB* and *AC* produced in the points *F* and *G* respectively, such that the line *GF* passes through *D*.

CG and *BF* are the required means.

Heron's construction

The diagram to show Heron's construction would be the same as that for Apollonius's method, but Heron in fact proceeds by making the line *GF* rotate about the point *D* and stops it when *EG* = *EF*. This brings out the analogy with Plato's method: as the line turns from the horizontal position, *DC*, to the vertical position *DB*, in the clockwise direction, *EG* increases from *EC* to + ∞ and *EF* decreases from + ∞ to *EB* = *EC*. Since the motion is continuous there will be one and only one position of the line for which *EG* = *EF*.

The construction of Philon of Byzantium

Another variant (Fig. 6.3): Philon draws the circle which circumscribes the rectangle. The straight line *GHDF* cuts

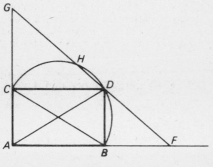

Fig. 6.3

the circle again in *H*. Philon stops the rotation of the line, or ruler, when *GH* = *DF*. This is a more convenient procedure to employ in practice.

Pappus's construction

Pappus himself says:

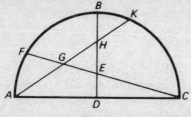

Fig. 6.4

We draw the semicircle *ABC* [Fig. 6.4]; from its centre *D* we construct the perpendicular *DB*, and move a ruler round the point *A* in such a manner that one of its ends is held by a pivot at *A* and the other end moves round between the points *B* and *C*. Having carried out these constructions let us now suppose that we are required to find two cubes which bear a given ratio to one another; we arrange that the given ratio shall be that of the straight line *BD* to the straight line *ED* and we produce the line joining *C* and *E* to meet the circle again at *F*. We now move the ruler between the points *B* and *C* until the length of it cut off between the straight lines *FE* and *EB* is equal to the length cut off between the straight line *BE* and the arc *BKC*. This can be done quite easily by making continual checks as the ruler is moved. Suppose it to have been done, and the ruler to be in the position *AGHK*, such that the straight lines *GH* and *HK* are equal; I say that the cube constructed on the straight line *DB* and the cube constructed on the straight line *DH* are in the required ratio, that is in the same ratio as the line *DB* and the line *DE*.

A little before Pappus, Sporus had given an almost identical construction.

Nicomedes's construction

We are given the two straight lines *CD* and *DA*, which are perpendicular to one another (Fig. 6.5).

We complete the rectangle *ABCD*. Let *L* be the mid-point of *AB* and *E* the mid-point of *BC*, and let *LD* cut *CB* produced in *G*. We take the point *F* on the perpendicular bisector of *BC* such that *FC* = *AL*. We join the points *G* and *F* and draw through *C* a line *CH* parallel to *GF*.

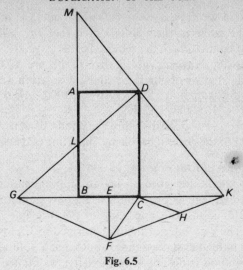

Fig. 6.5

We draw the straight line FHK such that $HK = AL = FC$. The straight line DK cuts AB produced in M. The required mean proportionals between DC and AD are CK and AM.

Huygens's construction

Huygens describes a construction similar to the ones we have just given but applicable only to the problem of doubling the cube.

We are required to construct a cube double that with side AB. We draw the circle $AFDC$ with centre B and radius AB (Fig. 6.6). We take a chord $AF = AB$ and join FC.

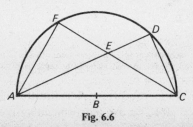

Fig. 6.6

Through A we draw a chord AD which cuts FC in E and meets the circle in the point D such that $DC = EF$. AE will then be the side of the required cube.

Moreover, Huygens notes that if the arc $CD = \pi/4$, although what we obtain is not an exact solution it is still a good <u>approximation</u>, for if $AB = 10\,000$, $AE \simeq 12\,605$ and $\sqrt{2 \times (10\,000)^3} \simeq 12\,599$.

Huygens also describes three methods of constructing solutions to the general problem of inserting two means.

The solution of Eratosthenes

All methods described so far are approximations, but they involve regular procedures using only monotonically varying functions and appealing to the principle of continuity. The very inferior method ascribed to Eratosthenes gives an unrelated succession of approximate solutions.

The method is described in the letter we quoted at the beginning of this chapter:

> Let there be given two unequal straight lines AE, HD between which it is required to find two mean proportionals in continued proportion [Fig. 6.7], and let AE be placed at right angles to the straight line EH, and upon EH let there be erected three successive parallelograms AF, FN, NH, and let the diagonals AF, MG, NH be drawn therein; these will be parallel. While the middle parallelogram FN remains stationary, let the other two approach each other, AE above the middle one, NH below it, as in the second figure, until A, B, C, D lie along a straight line, and let a straight line be drawn through the points A, B, C, D and let it meet EH produced in K ...
> ... Therefore between AE, DH two means, BF, CG, have been found.
> Such is the demonstration on geometrical surfaces; and in order that we may find the two means mechanically, a board of wood or ivory or bronze is pierced through, having on it three equal tablets, as smooth as possible, of which the midmost is fixed and the two outside run in grooves, their sizes and proportions being a matter of individual choice—for the proof is accomplished in the same manner; in order that the lines may be found with the greatest accuracy, the instrument must be skilfully made, so that when the tablets are moved everything remains parallel, smoothly fitting without a gap.

Eratosthenes's method is mathematically unsatisfactory in that the two tablets move independently of one another,

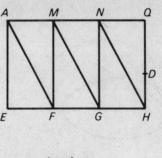

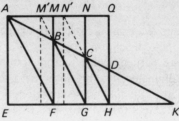

Fig. 6.7

so that each attempt at finding a solution is unrelated to its predecessor.

As well as these methods of continuous approximation there were also many methods which involved intersecting curves. Some of these methods are rather intuitive, but from a mathematical point of view they are just as satisfactory as the methods already described (apart from that of Eratosthenes). We shall begin with the earliest, and one of the most elegant, of the methods involving curves.

The construction of Archytas (fourth century BC)

Given two straight lines AC and AB, such that $AC > AB$, we are required to construct two mean proportionals

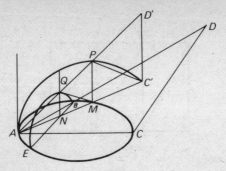

Fig. 6.8

between them. We draw a circle with diameter AC and the line AB as one of its chords (Fig. 6.8).

In a plane perpendicular to the first one we draw a circle with diameter AC and allow it to rotate about the perpendicular to the first plane at the point A. [The surface described is a half-torus with zero internal radius.] We construct the right cylinder whose base is the circle ABC. The cylinder will cut the surface of the torus in some fixed skew curve.

The straight line AB produced cuts the tangent at C to the circle ABC in the point D. We rotate the triangle ACD about the line AC. AD describes the surface of a cone of revolution. The skew curve we mentioned above cuts this cone in a point P. The plane through A and P perpendicular to the plane ABC cuts the torus in the circle APC', the cylinder in the vertical line PM and the cone in the straight line APD'. AM and AP are the required means.

If we construct a circle with the chord EB, parallel to CD, as diameter, the circle to lie in a plane perpendicular to ABC, APD', being the generator of a cone, will meet this circle in a point Q, and $AB = AQ$. Since A, Q, P and D' are collinear, the same is true of their orthogonal projections, the points A, N, M and C'. In the original circle $NB \times NE$

$= NA \times NM$. In the circle AQB $NQ^2 = BN \times NE$, therefore $NQ^2 = NA \times NM$ and angle AQN is a right angle, as is also the angle APC'.* We therefore have three similar triangles

$$AC'P, APM \text{ and } AMQ,$$

so

$$\frac{AC'}{AP} = \frac{AP}{AM} = \frac{AM}{AQ} \quad \text{or} \quad \frac{AM}{AB}.$$

Menaechmus's first construction

Let the given lines be at right angles and let $OA > AB$ (Fig. 6.9). Suppose the two means have been found to be ON and OM, then we have

$$\frac{OA}{OM} = \frac{OM}{ON} = \frac{ON}{OB}.$$

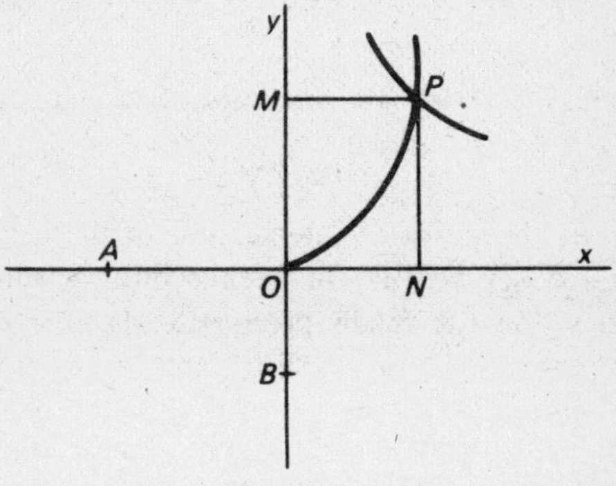

Fig. 6.9

---

* This argument is taken from Eutocius, and possibly derives from Eudemus, from whom Eutocius claims to have learned of this construction. The calculation is however quite unnecessary. The circle BQE obviously lies on the sphere which has great circle ABC, so the angle AQM is clearly a right angle. It may well be that Archytas realized this.

Therefore $OB.OM = ON^2 = PM^2$, so the point P (see figure) lies on a parabola ($OB.y = x^2$).

Also, $OA.OB = OM.ON = PN.PM$, so the point P lies on a hyperbola ($OA.OB = xy$).

P is therefore the point of intersection of these two curves, and the problem is thus solved.

Menaechmus's second construction

Given the same lines as before, we deduce that $OB.OM = ON^2 = PM^2$, and P is therefore on the same parabola as in the first solution; also, $OA.ON = OM^2 = PN^2$, therefore P is on another parabola ($OA.x = y^2$) (Fig. 6.10).

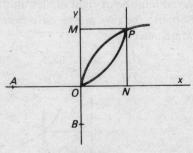

Fig. 6.10

P is thus the point of intersection of these two curves. It will be observed that in Menaechmus's solutions the given lines and the mean proportionals are arranged in exactly the same way as in Plato's mechanical solution.

Later, Diocles used the *cissoid* to solve this problem, obtaining a solution, rather like those of Sporus and Pappus.

Nicomedes's solution uses a *conchoid* to construct the segment $HK = AL$ proportional between the lines CK and CH.

We shall now turn to seventeenth-century solutions, and consider the work of Fermat, who, with Descartes, shows the beginnings of a new attitude to problems of this sort.

Fermat's justification of Menaechmus's constructions [4]

Suppose for example that we are required to find two mean proportionals.
Let the two lines between which we are to find the mean proportionals
be B and D, and let B be greater than D.

Let A be the greater of these means. We shall have A cubed equals B
squared into D [$a^3 = b^2d$]. Equate both these terms to B into A into
E [bae]. On one side we shall have A squared equals B into E [$a^2 = be$]
and on the other A into E equals B into D [$ae = bd$].

Therefore the problem will be solved by finding the point of intersection
of a hyberbola and a parabola.

Let the line OVN be in a fixed direction, and let the point 0 be on this
line. Given the straight lines B and D between which we are required to
find mean proportionals. Suppose OV is equal to A and let E be the straight
line VM perpendicular to OV. From the first equation A squared equals B
into E [$a^2 = be$], and it is clear that we must draw a parabola with O as
vertex, B as latus rectum, a diameter parallel to VM and ordinates parallel
to OV: this parabola will pass through the point M.

From the second equation:
B into D equals A into E [$bd = ae$]. Let us take any point N on the straight
line OV. Draw the line NZ perpendicular to OV at N and let the rectangle
contained by ON and NZ be equal to the rectangle B into D. Construct
the perpendicular OR. In accordance with our method of loci we are
required to draw a hyperbola passing through the point Z and having
asymptotes RO and OV; its position will be fixed and it will pass through
the point M.

But the parabola we mentioned earlier is also fixed and passes through
the same point M [Fig. 6.11]: therefore the point M is fixed. Therefore if

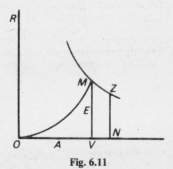

Fig. 6.11

we drop the perpendicular MV, the point V is given, and so therefore is the
line OV, which is the greater of the two mean proportionals we were
required to find.

The two mean proportionals can therefore be found from the point of intersection of a parabola and a hyperbola.

If we want to make the problem a quartic one we multiply all the terms by A:

A squared squared equals B squared into D into A $[a^4 = b^2da]$.

As in the previous method we equate each side of this equation to B squared into E squared $[b^2e^2]$, thus obtaining two equations, namely

A squared equals B into E $[a^2 = be]$

and D into A equals E squared $[da = e^2]$, and each equation will give a fixed parabola. The mean proportionals can therefore be found from the points of intersection of two parabolas.

These two constructions are described in Eutocius's commentary on the work of Archimedes; the above argument provides an immediate justification for them....

As an example of this construction* let us consider the problem of finding two means.

We have A cubed equals B squared into D $[a^3 = b^2d]$; therefore A squared squared equals B squared into D into A $[a^4 = b^2da]$. Add to each side

B squared squared $-$ B squared into A squared times two $[b^4 - 2b^2a^2]$:
A squared squared $+$ B squared squared $-$ B squared into A squared times two equals
B squared squared $+$ B squared into D into A $-$ B squared into A squared times two

$$[a^4 + b^4 - 2a^2b^2 = b^4 + b^2da - 2b^2a^2].$$

Let B squared times two equal N squared $[2b^2 = n^2]$ and equate each of these terms to N squared into E squared $[n^2e^2]$.

From one side we obtain

A squared $-$ B squared equals N into E $[a^2 - b^2 = ne]$; the end of E lies on a parabola. From the other side we have

B squared $\frac{1}{2}$ $+$ $D\frac{1}{2}$ into A $-$ A squared equals E squared

$$[\tfrac{1}{2}b^2 + \tfrac{1}{2}da - a^2 = e^2];$$

so the end of E will lie on a circle.

No attentive reader can fail to realise that the above problems cannot be solved in plane† terms, that is that the problems of mean proportionals, of trisecting an angle and other similar problems cannot be solved by the use of straight lines and circles alone.

---

* Fermat has just shown that any cubic or quartic equation can be solved by finding the points of intersection of a circle and a conic.

† i.e. by a construction using a straight edge and compasses.

6.2. Trisection of the angle

The origins of this problem are obscure. The problems of doubling the cube and squaring the circle led mathematicians to irrational or inexpressible ratios such as $\sqrt[3]{2}$ and π, but for the trisection of the angle the ratio concerned is a very simple one, expressible in numbers as $\frac{1}{3}$. It is a straightforward matter to trisect a right angle, since an angle of 30° can be constructed by means of a straight edge and compasses [a type of construction which the Greeks would have referred to as 'plane'], therefore, since an obtuse angle can be divided into a right angle plus an acute angle, the problem of trisecting an angle can be reduced, without loss of generality, to the problem of trisecting an acute angle.

One early solution to this problem is ascribed to Hippias of Elis, who lived in the fifth century BC.

To arrive at his solution Hippias used a curve called the *quadratrix*. (We shall explain the origin of this name in Chapter 7 (p. 200).)

The quadratrix is obtained by rotating a straight line *OM* about the centre of a quadrant of a circle *BOA*, from the position *OB* to the position *OA* (Fig. 6.12). The rotation

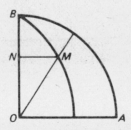

Fig. 6.12

is uniform. At the same moment that the line *OM* leaves the position *OB* a line *NM*, parallel to *OA*, leaves the point *B* and moves down uniformly so as to reach the position *OA* at the same time as the rotating line.

M, the point of intersection of the two lines, describes a quadratrix.

The quadratrix can be used to trisect an angle, or, indeed, to divide it in any ratio.

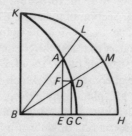

Fig. 6.13

Indeed, *writes Pappus*, let *LH* be an arc of the circle *KLH* [Fig. 6.13], and suppose that we are required to divide this arc in a given ratio [5].

We draw the straight lines *LB* and *BH* through the centre of the arc and draw the straight line *BK* at right angles to the line *BH*. We draw the quadratrix *KADC* through the point *K*, and on the perpendicular *AE* we mark the point *F* such that the ratio of *AF* to *AE* is the ratio in which we are required to divide the angle. We draw the straight line *FD* parallel to *BC* and then join the points *B* and *D* and draw the perpendicular *DG*. Now, from the property of the curve, the ratio of the angle (*AB, BC*) to the angle (*DB, BC*) is the same as the ratio of the line *AE* to the line *DG*, that is, to the line *FE*.

A quadratrix can be constructed point by point, by repeated bisections. The quadratrix constructed by finding three or four points in this way gives results comparable with those obtained with a cheap modern protractor.

The *Archimedean Spiral* can also be used to trisect an angle.

However, neither the quadratrix nor the Archimedean Spiral can be constructed exactly. Descartes would have expressed this fact by saying that neither of the curves was geometrical, since it is impossible to construct or even to design a mechanism which would coordinate the movements of rotation and translation in the manner required.

It is probable that as early as the third century BC Greek geometers tried to find other methods of trisecting the angle.

For example, we find as Proposition 8 of the *Lemmas* (ascribed to Archimedes):

If in a given circle we draw a general chord *AB* and produce it to the point *C* such that *BC* is equal to the radius of the circle, and draw the line *CD* (where *D* is the centre of the circle) and let this line cut the circle in *F* and *E*, then the arc *AE* is three times the arc *BF* [Fig. 6.14].

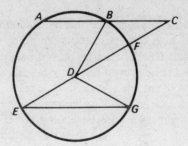

Fig. 6.14

For let us draw the line *EG* parallel to *AB*, and the lines joining *D* to *B* and *G*. Since the angle *DEG* is equal to the angle *DGE* the angle *GDC* is twice the angle *DEG*. Since the angle *BDC* is equal to the angle *ACE* the angle *GDC* is twice the angle *CDB*, and the total angle, angle *BDG*, is therefore three times the angle *BDC*, so the arc *BG*, which is equal to the arc *AE*, is three times the arc *BF*, which is what we were required to prove.

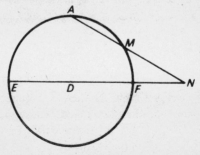

Fig. 6.15

To divide the circular arc *EA* into three parts we therefore first draw the diameter *EDF* (Fig. 6.15) and then rotate
about the point *A* a secant which cuts the circle in *M* and the
diameter produced in *N*. It is clear that as *N* goes from *F* to
infinity the length of the straight segment *MN* increases
from zero to infinity. This length will therefore once and only
once be equal to *DE*, the radius of the circle. When the
secant is in this position the arc *MF* will be one third of the
arc *EA*.

The construction is carried out by trial and error, but it
is easy to see how successive trials should be made: if *MN*
is greater than *DE*, *N* must be moved nearer to *F*, if it is
smaller, *N* must be moved further away: the dividers used
in the measurements do not need to be reset.

In practice, the method we have just described is thus
an improvement upon direct trial and error, which would
involve opening the dividers by some arbitrary amount,
measuring three times across the arc *EA*, and then re-
adjusting the dividers if their setting proved to be incorrect.

In Book IV of his *Collection* Pappus writes [6]:

... Since problems differ in this way, the earlier geometers were not
able to solve the aforementioned problem about the angle, when they
sought to do so by means of planes,* because it is by nature solid; for
they were not yet familiar with the sections of the cone, and for this reason
were at a loss. Later, however, they trisected the angle by means of the
conics, using in the solution the verging described below.

PROPOSITION 36. *Given a right-angled parallelogram ABCD, with BC
produced, let it be required to draw AE so as to make the straight line EF
equal to the given straight line* [Fig. 6.16].

The use of this verging (νεῦσις), that is interpolation,
is easy to justify: as *F* moves from *C* to infinity along *BC*
produced we can see that *EF* increases monotonically from
zero to infinity and therefore takes the given value once and
only once. Here too, as in Archimedes's lemma, we could
also use the *Conchoid* of Nicomedes.

_____
* Plane procedures are those which require only a straight edge and
compasses.

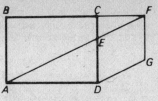

Fig. 6.16

But Pappus shows that the point G is the point of intersection of a hyperbola and a circle $[\overrightarrow{DG} = \overrightarrow{EF}]$.

DG is of given length and G therefore lies on a circle centre D. Also, DG is parallel to AEF therefore $FG = ED$ and $AB \times BC = BF \times FG$.

G therefore lies on an equilateral hyperbola passing through D and having asymptotes BA and BCF.

The interpolation in question therefore resolves itself into a 'solid problem'.

PROPOSITION 38. *With this proved, the given rectilineal angle is trisected in the following manner.*

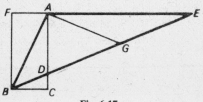

Fig. 6.17

First let ABC be an acute angle [Fig. 6.17], and from any point [of the straight line AB] let the perpendicular AC be drawn, and let the parallelogram CF be completed, and let FA be produced to E, and inasmuch as CF is a right-angled parallelogram let the straight line ED be placed between EA, AC so as to verge towards B and be equal to twice AB— that this is possible has been proved above; I say that EBC is a third part of the given angle AB.

PROPOSITION 43. *Another way of cutting off the third part of a given arc is furnished, without the use of a verging, by this solid locus.*

Let the straight line through A, C be given in position, and from the given points A, C upon it let ABC be inflected, making the angle ACB double of CAB; I say that B lies on a hyperbola [Fig. 6.18].

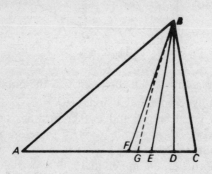

Fig. 6.18

For let BD be drawn perpendicular [to AC] and let DE be cut off equal to CD; when BE is joined it will therefore be equal to AE.* And let EF be placed equal to DE; therefore $CF = CD$. Now let CG be placed equal to $\frac{1}{3}AC$; therefore the point G will be given, and the remainder† AF will equal $3GD$.

Now since‡ $BE^2 - EF^2 = BD^2$ and $BE^2 - EF^2 = DA . AF$, therefore

$$DA . AF = BD^2$$

that is

$$3AD . DG = BD^2$$

---

* For by the equality of the triangles BED, BCD, we have $\angle BEC = \angle BCE = 2\angle CAB$ (ex hypothesi). But $\angle BEC = \angle CAB + \angle ABE$. Therefore $\angle CAB = \angle ABE$, and so $BE = AE$.

† i.e. since $CG = \frac{1}{3}AC$ and $CD = \frac{1}{3}CF$, by subtraction, $CG - CD = \frac{1}{3} \times (AC - CF)$, or $GD = \frac{1}{3}AF$.

‡ The reasoning here is much abbreviated, and in full may be written as follows:

$$BE^2 - EF^2 = BE^2 - ED^2 \text{ (since } EF = ED \text{ ex hypothesi)}$$

$$= BD^2 \qquad \text{[Euclid I, 47]}$$

(Cont.)

therefore B lies on a hyperbola with transverse axis AG and conjugate axis $\sqrt{3}AG$*....

And the synthesis is clear; for it will be required so to cut AC that AG is double of GC, and about AG as axis to describe through G hyperbola with conjugate axis $\sqrt{3}AG$, and to prove that it makes the aforementioned double ratio of the angles. And that the hyperbola described in this manner cuts off the third part of the arc of the given circle is easily understood if the points A, C are the end points of the arc.†

PROPOSITION 44. *Some set out differently the analysis of the problem of trisecting an angle or arc without a verging. Let the ratio be upon an arc; it makes no difference whether an angle or an arc is to be divided* [Fig. 6.19].

Let it be done, and let BC, the third part of the arc ABC, be cut off, and let AB, BC, CA be joined, then $\angle ACB = 2\angle BAC$. Let $\angle ACB$ be bisected by CD, and let DE, FB be drawn perpendicular; therefore AD is equal to DC, so that AE is also equal to EC; therefore E is given.

---

* This means that the transverse axis of the hyperbola is AG [where A and G are the vertices of the hyperbola] and that the ratio of the two axes is $\sqrt{3}$. See page 92: the square of DB is applied to AG in hyperbola by a rectangle similar to that with base AG (transverse axis) and height $\sqrt{3}AG$ (conjugate axis, latus rectum). This last rectangle is 'the figure applied along the axis'.

† *Pappus* leaves it as an exercise to the reader to use this result to trisect a circular arc.

Now

$$BE^2 - EF^2 = AE^2 - EF^2 \text{ (since } BE \text{ was proved equal to } AE)$$

$$= DA \cdot AF \text{ [Euclid II, 6]}$$

$$\therefore \quad DA \cdot AF = BD^2$$

$$\therefore \quad 3AD \cdot DG = BD^2 \text{ (since } AF \text{ was proved equal to } 3GD)$$

$$\therefore \quad BD^2 : AD \cdot DG = 3 : 1$$

$$= \frac{3AG^2}{AG^2};$$

$\therefore B$ lies on a hyperbola with transverse axis AG and conjugate axis $\sqrt{3}AG$.

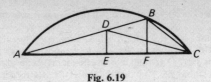

Fig. 6.19

Now because

$$AC:CB = AD:DB \text{ [Euclid VI, 5]}$$

$$= AE:EF.$$

therefore alternatively $CA:AE = BC:EF$. But $CA = 2AE$; and therefore $BC = 2EF$; therefore $BC^2 = 4EF^2$, that is, $BF^2 + FC^2 = 4EF^2$. Now, since the two points E, C are given, and BF is drawn at right angles, and the ratio $EF^2:BF^2 + FC^2$ is given, B lies on a hyperbola.* But it also lies on an arc given in position; therefore B is given. And the synthesis is clear.

The problem of trisecting an angle, like that of doubling the cube, excited great interest among Western mathematicians in the sixteenth and seventeenth centuries.

In the sixteenth century, Tartaglia and Cardan reduced the problem of solving certain cubic equations to that of finding two geometrical means (see Vol. I, Chapter 7).

We shall describe the principle of their method as it is given by Hudde, since Tartaglia and Cardan's own work is very difficult to read. (Their equations are purely numerical and are written out in full, using hardly any conventional

---

* Let $EC = a$, $EF = x$, $BF = y$. Then we have

$$y^2 + (a - x)^2 = 4x^2 \quad \text{or} \quad y^2 = 3x^2 + 2ax - a^2,$$

$$y^2 = 3\left(x + \frac{a}{3}\right)^2 - \frac{4a^2}{3}, \quad \text{i.e.,} \quad 3\left(x + \frac{a}{3}\right)^2 - y^2 = \frac{4a^2}{3},$$

and by moving the axis we can transform this into the canonical form

$$\frac{x^2}{m^2} - \frac{y^2}{n^2} = 1.$$

symbols: and since negative numbers are not employed many different cases have to be considered.)

Let the equation be

$$x^3 + ax^2 + bx + c = 0.$$

Writing $x = y - a/3$ we obtain

$$\left(y - \frac{a}{3}\right)^3 + a\left(y - \frac{a}{3}\right)^2 + b\left(y - \frac{a}{3}\right) + c = 0,$$

i.e.

$$\left(y^3 - ay^2 + \frac{a^2}{3}y + \frac{a^3}{27}\right) + ay^2 - \frac{2a^2y}{3} + \cdots = 0,$$

that is, an equation of the form

$$y^3 + py + q = 0,$$

which contains no term in y^2.

Cardan, Vieta and Descartes all explain how to rewrite the given equation in this form.

We shall follow Hudde, writing $u + v$ for y, and noting that $(u + v)^3 = u^3 + v^3 + 3uv(u + v)$. The equation can now be written

$$u^3 + v^3 + q + (u + v)[3uv + p] = 0.$$

Since we have two unknowns, u and v, we may set both $u^3 + v^3 + q$ and $3uv + p$ separately equal to zero.

Then we have $u^3 + v^3 = -q$ and $uv = -p/3$ i.e. $u^3v^3 = -p^3/27$.

Since we know the sum and the product of u^3 and v^3 we can find u^3 and v^3 themselves, and then, by taking cube roots, obtain u and v and thus y, which is equal to $u + v$.

This method can only be used if the values of u^3 and v^3 are real, that is, if the equation $X^2 + qX - p^3/27 = 0$, of which u^3 and v^3 are the roots, has a non-negative discriminant, i.e. if $27q^2 + 4p^3 \geq 0$.

If this condition is not satisfied, Cardan's method cannot be used, and solving the equation is not equivalent to calculating geometric means. In this case, the equation has three roots, and Vieta showed that solving it is equivalent to trisecting an angle.

This point is an important one, but we cannot discuss it here.

Descartes deals with the problem of trisecting an angle in Book III of his *Geometry (Géométrie)*. He begins by finding an equation to express the problem (a process which Vieta would have called zetetic analysis). He then solves the equation graphically (a process which Vieta would have called rhetic or exegetic analysis) by considering the points of intersection of a circle and a parabola. Descartes justified this procedure a few pages earlier by general considerations.

If we are required to divide the angle *NOP* [Fig. 6.20] or the circular arc *NQTP* into three equal parts: by setting *NO* ∞ 1* for the radius of the circle, and *NP* ∞ *q* for the length of the chord subtending the given arc, and *NQ* ∞ *z* for the chord subtending one third of this arc, we obtain the equation

$$z^3 \infty 3z - q.$$

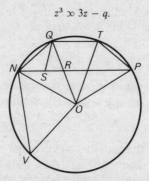

Fig. 6.20

For if we draw the lines *NQ*, *OQ* and *OT*, and the line *QS*, parallel to *TO*, we can see that since the ratio of *NO* to *NQ* is the same as the ratio of *NQ* to *QR* and the ratio of *QR* to *RS*, so if we make *NO* equal to 1 and

---

* Descartes uses ∞ to signify 'equals'.

NQ equal to z then QR is z^2 and RS is z^3; and since it is only by the length RS, or z^3, that the line NP, which is of length q, is less than three times NQ, which is z, we have $q \propto 3z - z^3$ or $z^3 \propto 3z - q$.*

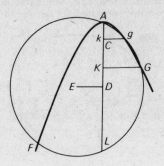

Fig. 6.21

Now, if we draw the parabola FAG [Fig. 6.21], and set CA, its semi-latus rectum, as $\frac{1}{2}$ and take $CD \propto \frac{3}{2}$ and the perpendicular $DE \propto \frac{1}{2}q$ and draw a circle $FAgG$, with centre E and radius EA, then this circle cuts the parabola at the three points F, g and G as well as at the point A, the vertex of the parabola. This shows that the equation has three roots, that is, two true roots GK and gk, and a third false one FL.† And of these two true roots it is the smaller, gk, which we must take as the required line NQ; for the other root, GK, is equal to NV, the chord subtending one third of the arc NVP, which together with the other arc, NQP, makes up the whole circle. And the false root, FL, is equal to the sum of these two, QN and NV, as is clear from the calculation.

Roberval gives a quite different solution, first proposed by Etienne Pascal. This solution involves the use of a new curve, Pascal's limaçon, one of the many curves which can be used for trisection. ('*On the Limaçon of M.P.*')‡

---

* We obtain this result by considering the similar isosceles triangles ONQ, NQR and QRS. J.V.F.

† True root = positive root; false root = negative root. Descartes has already shown that any cubic or quartic equation can be solved by finding the points of intersection of a circle with a parabola.

‡ Works of Roberval, 'On composite movements' ('Des mouvements composés'), lessons drawn up by du Verdus, a pupil of Roberval, about 1640.

The limaçon is another type of circular conchoid, which is described as follows:

Given the circle *CGBE* [Fig. 6.22], with centre *A*, and diameter *BC*, which may be indefinitely produced, for instance to the point *D*, let us

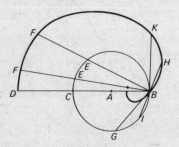

Fig. 6.22

take the point *B* as the pole of the Limaçon, and *CD* as the length required to describe it, a length which is less than the diameter of the circle. From *B* we draw a number of lines *BEF* across the circle, to cut its circumference in the points *E*, and on each of these lines we take the distance *EF*, equal to *CD*, and on the same side [of the circle]. The Limaçon will pass through all these points *FF*. Now we must note that we take as many intervals as we can, beginning with the convex part of the circle, the part on the same side of the line *BCD* as the Limaçon is, and then, to continue the curve, we must consider points *E* on the other half of the circumference of the circle, the part which is concave towards the Limaçon. Thus, the point *B* of the Limaçon corresponds to the point *G* on the circumference of the circle, where *BG* is equal to *CD*; and the last point of the Limaçon, which we have marked with a little *, corresponds to the point *C*, and the points of the Limaçon between *B* and * correspond to the points on the circumference of the circle from *G* to *C*. Similarly, the points of the Limaçon which lie closest to *B* above the diameter *CB* correspond to the points on the circumference of the circle from *G* to *B*, from *H*, which corresponds to *I*, round to the point *K*, which corresponds to *B*, and you can see from this why we noted that the length *CD* must not be greater than the diameter *CB* . . .''

[Roberval notes, however, that this restriction does not apply when we merely require the tangents to the curve. See Vol. 1, p. 247].

I shall mention in passing that if we take the length *DC* equal to *CA*, the semi-diameter of the circle about which the Limaçon is described, the

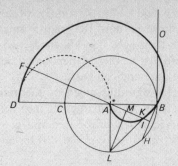

Fig. 6.23

smaller part of the Limaçon *B will divide a given rectilineal angle into three equal parts. This property was discovered by Pascal.†

For suppose we are given the angle DBH [Fig. 6.23], on one of the lines enclosing it, say the line DB, I take the point *, and from *I drop a perpendicular *I onto the other line, BH and this perpendicular cuts the arc *KB of the Limaçon (described with pole B round the circle with centre *, and radius *B and such that the standard length of the Limaçon, CD, is equal to *B) in the point K, I then draw the line BKL, and I assert that the angle between this line and the line BH, i.e. the angle KBH, is ⅓ of the given angle CBH.

To prove this, let us draw the circle of the Limaçon and produce the line BK to meet the circumference of the said circle in L, we then draw L*, and having divided K* *bifariam*‡ at M, draw the line LM, which will be perpendicular to AK; for, from the definition of the Limaçon, the sides L* and LK of the triangle *LK are equal to one another, both being equal to the same length, CD. Then, since LMK and BIK are right-angled triangles, and also each contain one of a pair of angles opposed at their vertex, these triangles are similar, and angle MLK is equal to angle BIK, but MLK is half of angle *LK (since the triangle *LK is isosceles and its base *K is divided *bif.* etc.), that is, half of the angle *BL (since the triangle *LB is also isosceles), and hence the angle KBH is half the angle *BL, and hence ⅓ of the whole angle *BH; which is what we were required to prove.

These examples suffice to show that in the sixteenth century the two very ancient problems of doubling the cube and of trisecting the angle were seen to be of great importance. Vieta identified them as the two problems

---

† i.e. Etienne Pascal, Blaise Pascal's father.
‡ Into two equal parts.

which if solved would lead to the solution of cubic equations.

It was not, however, until 1837 that Wantzel* (a French mathematician then aged 23) laid those two problems to rest by showing that neither of them could be solved using only a straight edge and compasses.

Pure geometry was thus aided by theoretical work on algebraic equations, which had come into being 250 years earlier, when Vieta had introduced the use of algebraic symbols (the *literal calculus* or *logistica speciosa*).

---

* 'Investigations of means of discovering whether a geometrical problem can be solved by means of a straight edge and compasses' ('Recherches sur les moyens de reconnaître si un problème de géométrie peut se résoudre par la règle et le compas') (*Journal de Mathématiques*, Volume II, 1837).

7. Squaring the Circle

*De la proportion du dyametre a sa ligne circonferentiale un sage homme tient que c'est comme de 7 a 22. Mais c'est chose qui ne se peut prouver par aucune demonstration.**

Nicolas Chuquet, 1484

The phrase 'squaring the circle' is commonly used to describe two closely related but nevertheless distinct problems. The first is that of finding a square which has the same area as a given circle. This is the true problem of squaring the circle.

In fact, we may be required to *calculate* the area of a circle of given radius and then to calculate the side of the equivalent square, or we may be required to *construct* this square directly. The construction can, or must, be carried out by means of some given instrument, and the result of the calculation, or the figure constructed, must be accurate to some degree which is either stated explicitly or tacitly assumed.

The second problem described as 'squaring the circle' is that of rectifying the circumference of a circle: we are required to find, either by calculation or construction, the length of a straight segment having a length equal to that of the circumference of the circle.

We shall give a brief account of the solutions proposed to these problems over a period of nearly four thousand years.

---

* The learned take the ratio of the diameter to the circumference to be 7 to 22. But this cannot be proved.

The oldest surviving document which refers to the problem of squaring the circle is the Rhind papyrus,* which was written in about 1800 BC by a scribe named Ahmes, who was copying from a document which was then already one or two centuries old.

Ahmes finds the area of a circle with diameter d as follows: he takes $\frac{1}{9}d$ away from d, multiplies the result by d, and then subtracts one ninth of the product from the product itself, i.e. he calculates $(1 - \frac{1}{9})d \times d - \frac{1}{9}(1 - \frac{1}{9})d \times d$ which is $(\frac{8}{9}d)^2$. This makes the circle equivalent to a square whose side is $\frac{8}{9}$ of the diameter of the circle.

Let us consider the accuracy of this value. The Egyptians used only unit fractions (i.e. fractions with numerator 1) in their calculations, so we want to know which such fraction best expresses the side of the equivalent square in terms of the diameter of the circle, i.e. which of the *subepimoric* ratios (1 minus a unit fraction) is the closest to the ratio of the side of the square to the diameter of the circle. If the side of the square is a and the diameter of the circle is d then we have $a^2 = \pi(d^2/4)$, therefore $a/d = \frac{1}{2}\sqrt{\pi}$. The value of $\sqrt{\pi}$ is $1 \cdot 77245 \ldots$, so $a/d = 0.8862 \ldots$.

Let us calculate the various subepimoric ratios:

$$1 - \tfrac{1}{7} = \tfrac{6}{7} = 0 \cdot 857 \ldots \qquad 1 - \tfrac{1}{8} = \tfrac{7}{8} = 0 \cdot 875$$

$$1 - \tfrac{1}{9} = \tfrac{8}{9} = 0 \cdot 8888 \ldots \qquad 1 - \tfrac{1}{10} = \tfrac{9}{10} = 0 \cdot 9.$$

The best approximation is undoubtedly $\frac{8}{9}$, the value chosen by the Egyptians. It might therefore be said that they had succeeded in squaring the circle.

We have no idea how this very satisfactory result was obtained. Geometry was still at a rudimentary stage and the Egyptians unhesitatingly assumed that the ratio in

---

* Now in the British Museum. J.V.F.

question remained the same from circle to circle: if one circle could be squared so could all circles. Thus some experimental procedure or intuitive geometrical process, even if rather crude, might in fact provide a satisfactorily accurate result. We know of good pairs of scales dating back to 5000 BC, and by about 2000 BC scales had developed to the point where the required ratio could have been found by weighing. Two vases could have been made, one with a square base and one with a circular base equal in size to the circle inscribed in the square, both vases being of the same height. The ratio of the weights of the volumes of water contained by these two vases would give the required result.

Some historians put forward a different suggestion, to which we shall now turn.

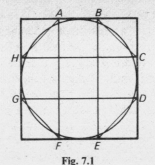

Fig. 7.1

Consider a square, and a circle inscribed in it (Fig. 7.1). We divide each side of the square into three equal parts, and by cutting off the four corners of the square at 45° we construct the semi-regular octagon $ABCDEFGH$. This octagon appears in many different places and at many different dates in the patterns of paving stones and tiles. We can regard the octagon as roughly equivalent to the inscribed circle, and its area is clearly $\frac{7}{9}$ of the area of the original square. so the

equivalent square would have a side $\sqrt{\frac{7}{9}}$ times that of the
original square.

Let us now take the square root of $\frac{7}{9}$, i.e. of $1 - \frac{2}{9}$. Using
known methods (see pages 66 to 71) we should obtain as
a first approximation the value $1 - \frac{1}{9}$, i.e. $\frac{8}{9}$.

At first sight this suggestion seems an attractive one, but
it has one weak point: we are in effect taking π to be
$\frac{7}{9} \times 4 = 3\frac{1}{9}$, and the best of the values of π calculated using
unit fractions is that of Archimedes, who obtained $3\frac{1}{7}$.
The value of π actually used by the Egyptians was $4 \times (\frac{8}{9})^2 =
\frac{256}{81} = 3\frac{13}{81}$, which lies between $3\frac{1}{7}$ and $3\frac{1}{6}$, and is thus much
more accurate than the value from which we are supposing
it to have been deduced.

It had long been thought that the Babylonians used the
crude value of 3 for π, but in about 1950 E. M. Bruins
deciphered some tablets found at Susa, which proved that
this was not the case.

To find the area of a circle the Babylonians first multiplied
the circumference by 5, or divided it by 12, which is an
equivalent operation in sexagesimal arithmetic. The result
was then refined by multiplying it by the factor expressed in
sexagesimal numbers as 57,36. Now, we know that if the circumference of a circle is C and the area of the circle is A
we have $A = (1/4\pi)C^2$, so the process we have just described
is equivalent to setting $1/4\pi = 5 \times 57,36$.

The reciprocal of 57,36 is 1;2,30 and the reciprocal of 5
is 12 so we have

$$4\pi = 12 \times 1;2,30,$$

i.e. $$\pi = 3 \times 1;2,30 = 3;7,30,$$
or

$$3 + \frac{7}{60} + \frac{1}{120} = 3\frac{15}{120} = 3\frac{1}{8}.$$

This same approximation is found in the Indian
Sulvasutras (dating from between 400 and 200 BC) where it

is stated that a square is equivalent to a circle if the diameter of the circle is $\frac{8}{10}$ of the diagonal of the square. The same approximation appears again, at the end of the fifteenth century and the beginning of the sixteenth, in the work of Dürer and of Charles de Bouvelles, and Tartaglia mentions it as useful for constructing solutions.

So by the beginning of the second millenium BC two very accurate approximate values of π were in use. The Egyptians used $3\frac{13}{81} \simeq 3\cdot1604$, a value which is too large by about $0\cdot0188$; and the Babylonians used $3\frac{1}{8} \simeq 3\cdot125$, a value which is too small by about 0.01666.

We shall see later how Archimedes derived, or justified an intermediate value of $3\frac{1}{7}$, a value which was then adopted by others.

We should note that both the Egyptians and the Babylonians assumed that the area of a circle was proportional to the area of the circumscribing square. The Greeks seem to have been the first people to realise that this needed to be proved.

Among the mathematicians of the fifth and fourth centuries who were concerned with the problem of squaring the circle we find Anaxagoras (500–428 BC), who seems to have worked on the problem while he was in prison, and Antiphon and Bryson, whose work, which is criticized by Aristotle, seems to have involved crude methods somewhat similar to those employed by Euclid and Archimedes. Hippocrates of Chios also addressed himself to the problem at this time, using a method of his own which we shall discuss later.

The surviving text of Archimedes's short treatise *Measurement of a circle*, which is probably not quite complete, consists of three propositions, the last and most important one being divided into two parts. In Proposition 1 Archimedes proves that the problems of squaring the circle and of rectifying its circumference are equivalent. In Proposition

2 he shows that if we assume that the circumference of a circle is three and one seventh times as long as its diameter, then the area of the circle is $\frac{11}{14}$ of the area of the circumscribing square. In Proposition 3 he first shows that the circumference of a circle is less than $3\frac{1}{7}$ times the diameter and then goes on to show that it is greater than $3\frac{10}{71}$ times the diameter.

PROPOSITION 1[1]. *The area of any circle is equal to a right-angled triangle in which one of the sides about the right angle is equal to the radius, and the other to the circumference, of the circle.*

Let ˉABCD be the given circle, K the triangle described [Fig. 7.2].

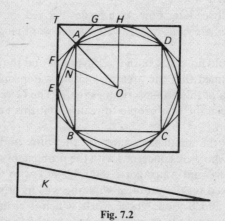

Fig. 7.2

Then, if the circle is not equal to K, it must be either greater or less.

I. If possible, let the circle be greater than K.

Inscribe a square ABCD, bisect the arcs AB, BC, CD, DA, then bisect (if necessary) the halves, and so on, until the sides of the inscribed polygon whose angular points are the points of division subtend segments whose sum is less than the excess of the area of the circle over K.

Thus the area of the polygon is greater than K.

Let AE be any side of it, and ON the perpendicular on AE from the centre O.

Then ON is less than the radius of the circle and therefore less than one of the sides about the right angle in K. Also the perimeter of the polygon is less than the circumference of the circle, i.e. less than the other side about the right angle in K.

Therefore the area of the polygon is less than K; which is inconsistent with the hypothesis.

Thus the area of the circle is not greater than K.

II. If possible, let the circle be less than K.

Circumscribe a square, and let two adjacent sides, touching the circle in E, H, meet in T. Bisect the arcs between adjacent points of contact and draw the tangents at the points of bisection. Let A be the middle point of the arc EH, and FAG the tangent at A.

Then the angle TAG is a right angle. Therefore

$$TG > GA$$

$$> GH.$$

It follows that the triangle FTG is greater than half the area $TEAH$.

Similarly, if the arc AH be bisected and the tangent at the point of bisection be drawn, it will cut off from the area GAH more than one-half.

Thus, by continuing the process, we shall ultimately arrive at a circumscribed polygon such that the spaces intercepted between it and the circle are together less than the excess of K over the area of the circle.

Thus the area of the polygon will be less than K.

Now, since the perpendicular from O on any side of the polygon is equal to the radius of the circle, while the perimeter of the polygon is greater than the circumference of the circle, it follows that the area of the polygon is greater than the triangle K; which is impossible.

Therefore the area of the circle is not less than K.

Since then the area of the circle is neither greater nor less than K, it is equal to it.

PROPOSITION 2. *The area of a circle is to the square on its diameter as* 11 *to* 14.

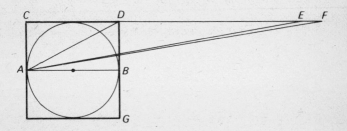

Given a circle of diameter AB, we circumscribe around it the square CG. Let DE be twice CD, and EF one-seventh of CD (see figure). Since the ratio of ACE to ACD is 21 to 7, and the ratio of ACD to AEF is 7 to 1,

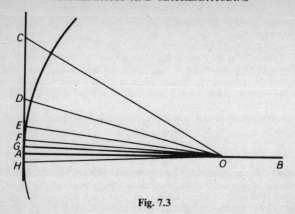

Fig. 7.3

the ratio of ACF to ACD is 22 to 7. But the square CG is four times the triangle ACD, and the triangle ACE is equal to the circle AB. Thus the circle is to the square CG the same ratio as 11 to 14.*

PROPOSITION 3. *The ratio of the circumference of any circle to its diameter is less than $3\frac{1}{7}$ but greater than $3\frac{10}{71}$.*

I. Let AB be the diameter of any circle, O its centre, AC the tangent at A; and let the angle AOC be one-third of a right angle [Fig. 7.3].

Then

$$OA : AC[= \sqrt{3} : 1] > 265 : 153 \qquad (1)$$

and

$$OC : CA[= 2 : 1] = 306 : 153 \qquad (2)$$

First, draw OD bisecting the angle AOC and meeting AC in D.

Now

$$CO : OA = CD : DA \text{ [Eucl. VI. 3],}$$

so that

$$[CO + OA : OA = CA : DA, \text{ or]}$$

$$CO + OA : CA = OA : AD.$$

Therefore [by (1) and (2)]

$$OA : AD > 571 : 153 \qquad (3)$$

---

* The proposition is translated from the French. T. L. Heath did not translate it. He writes: "The text of this proposition is not satisfactory, and Archimedes cannot have placed it before Proposition 3, as the approximation depends upon the result of that proposition." J.V.F.

Hence

$$OD^2 : AD^2 [= (OA^2 + AD^2) : AD^2$$

$$> (571^2 + 153^2) : 153^2]$$

$$> 349\,450 : 23\,409,$$

so that

$$OD : DA > 591\tfrac{1}{8} : 153. \tag{4}$$

Secondly, let OE bisect the angle AOD, meeting AD in E.
[Then

$$DO : OA = DE : EA,$$

so that

$$DO + OA : DA = OA : AE.]$$

Therefore

$$OA : AE [> (591\tfrac{1}{8} + 571) : 153, \text{ by (3) and (4)]}$$

$$> 1162\tfrac{1}{8} : 153. \tag{5}$$

[It follows that

$$OE^2 : EA^2 > \{(1162\tfrac{1}{8})^2 + 153^2\} : 153^2$$

$$> (1350534\tfrac{33}{64} + 23409) : 23409$$

$$> 1373943\tfrac{33}{64} : 23409.]$$

$$OE : EA > 1172\tfrac{1}{8} : 153 \tag{6}$$

Thirdly, let OF bisect the angle AOE and meet AE in F.

We thus obtain the result [corresponding to (3) and (5) above] that

$$OA : AF [> (1162\tfrac{1}{8} + 1172\tfrac{1}{8}) : 153]$$

$$> 2334\tfrac{1}{4} : 153. \tag{7}$$

Therefore

$$OF^2 : FA^2 > \{(2334\tfrac{1}{4})^2 + 153^2\} : 153^2$$

$$> 5472132\tfrac{1}{16} : 23409$$

Thus

$$OF : FA > 2339\tfrac{1}{4} : 153. \tag{8}$$

Fourthly, let OG bisect the angle AOF, meeting AF in G.
We have then

$$OA : AG [> (2334\tfrac{1}{4} + 2339\tfrac{1}{4}) : 153, \text{ by means of (7) and (8)]}$$

$$> 4673\tfrac{1}{2} : 153.$$

Now the angle AOC, which is one-third of a right angle, has been bisected four times, and it follows that

$$\angle AOG = \tfrac{1}{48} \text{ (a right angle)}.$$

Make the angle AOH on the other side of OA equal to the angle AOG, and let GA produced meet OH in H.

Then

$$\angle\,GOH = \tfrac{1}{24}\text{ (a right angle).}$$

Thus GH is one side of a regular polygon of 96 sides circumscribed to the given circle.

And, since

$$OA:AG > 4673\tfrac{1}{2}:153,$$

while

$$AB = 2OA, \qquad GH = 2AG,$$

it follows that

$$AB:\text{(perimeter of polygon of 96 sides) } [> 4673\tfrac{1}{2}:153 \times 96]$$

$$> 4673\tfrac{1}{2}:14688.$$

But

$$\frac{14\,688}{4673\tfrac{1}{2}} = 3 + \frac{667\tfrac{1}{2}}{4673\tfrac{1}{2}}$$

$$\left[< 3 + \frac{667\tfrac{1}{2}}{4672\tfrac{1}{2}}\right].$$

$$< 3\tfrac{1}{7}.$$

Therefore the circumference of the circle (being less than the perimeter of the polygon) is *a fortiori* less than $3\tfrac{1}{7}$ times the diameter AB.

II. Next let AB be the diameter of a circle, and let AC, meeting the circle in C, make the angle CAB equal to one-third of a right angle. Join BC [Fig. 7.4].

Then

$$AC:CB[= \sqrt{3}:1]\quad 1351:780.$$

First, let AD bisect the angle BAC and meet BC in d and the circle in D. Join BD.

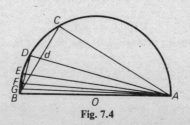

Fig. 7.4

Then
$$\angle BAD = \angle dAC$$
$$= \angle dBD,$$
and the angles at D, C are both right angles.

It follows that the triangles ADB, $[ACd]$, BDd are similar.

Therefore
$$AD:DB = BD:Dd$$
$$[= AC:Cd]$$
$$= AB:Bd \text{ [Eucl. VI, 3]}$$
$$= AB + AC:Bd + Cd$$
$$= AB + AC:BC$$

or
$$BA + AC:BC = AD:DB.$$

[But
$$AC:CB < 1351:780, \text{ from above,}$$

while
$$BA:BC = 2:1$$
$$= 1560:780.]$$

Therefore
$$AD:DB < 2911:780. \qquad (1)$$

[Hence
$$AB^2:BD^2 < (2911^2 + 780^2):780^2$$
$$< 9\,082\,321:608\,400.]$$

Thus
$$AB:BD < 3013\tfrac{3}{4}:780. \qquad (2)$$

Secondly, let AE bisect the angle BAD, meeting the circle in E; and let BE be joined.

Then we prove, in the same way as before, that

$$AE:EB[= BA + AD:BD$$
$$< (3013\tfrac{3}{4} + 2911):780, \text{ by (1) and (2)]}$$
$$< 5924\tfrac{3}{4}:780$$
$$< 5924\tfrac{3}{4} \times \tfrac{4}{13}:780 \times \tfrac{4}{13}$$
$$< 1823:240. \qquad (3)$$

[Hence

$$AB^2 : BE^2 < (1823^2 + 240^2) : 240^2$$

$$< 3\,380\,929 : 57\,600.]$$

Therefore

$$AB : BE < 1838\tfrac{9}{11} : 240. \tag{4}$$

Thirdly, let AF bisect the angle BAE, meeting the circle in F.
Thus

$$AF : FB[= BA + AE : BE$$

$$< 3661\tfrac{9}{11} : 240, \text{ by (3) and (4)}]$$

$$< 3661\tfrac{9}{11} \times \tfrac{11}{40} : 240 \times \tfrac{11}{40}$$

$$< 1007 : 66. \tag{5}$$

[It follows that

$$AB^2 : BF^2 < (1007^2 + 66^2) : 66^2$$

$$< 1018405 : 4356.]$$

Therefore

$$AB : BF < 1009\tfrac{1}{6} : 66. \tag{6}$$

Fourthly, let the angle BAF be bisected by AG meeting the circle in G.
Then

$$AG : GB[= BA + AF : BF]$$

$$< 2016\tfrac{1}{6} : 66, \text{ by (5) and (6)}.$$

[And

$$AB^2 : BG^2 < \{(2016\tfrac{1}{6})^2 + 66^2\} : 66^2$$

$$< 4\,069\,284\tfrac{1}{36} : 4356.]$$

Therefore

$$AB : BG < 2017\tfrac{1}{4} : 66,$$

whence

$$BG : AB > 66 : 2017\tfrac{1}{4}. \tag{7}$$

[Now the angle BAG which is the result of the fourth bisection of the angle BAC, or of one-third of a right angle, is equal to one-fortyeighth of a right angle.

Thus the angle subtended by *BG* at the centre is

$$\tfrac{1}{24} \text{ (a right angle).]}$$

Therefore *BG* is a side of a regular inscribed polygon of 96 sides. It follows from (7) that

$$(\text{perimeter of polygon}) : AB[> 96 \times 66 : 2017\tfrac{1}{4}]$$

$$> 6336 : 2017\tfrac{1}{4}.$$

And

$$\frac{6336}{2017\tfrac{1}{4}} > 3\tfrac{10}{71}.$$

Much more then is the circumference of the circle greater than $3\tfrac{10}{71}$ times the diameter.

Thus the ratio of the circumference to the diameter
$$< 3\tfrac{1}{7} \text{ but } > 3\tfrac{10}{71}.$$

The original text of the third proposition is not easy to read: the argument is simple and elegant but no working is shown (only the results of the calculations are given) and this makes it difficult to follow the line of Archimedes's reasoning.

Since this proposition is in fact a fine example of geodesy (practical geometry) and of logistics (numerical calculation) we shall attempt to elucidate it. The calculations refer to the extraction of square roots. We refer the reader to Chapter 3 for an explanation of how this process was carried out.

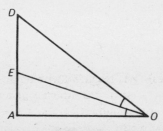

Fig. 7.5

The geometrical part is based on Pythagoras's Theorem, on results to do with the bisector of an angle of a triangle (Euclid, *Elements*, Book VI, Proposition 3) and, in the second part, refers to a corollary to Euclid's proposition.

First part (Fig. 7.5). The following argument is used repeatedly: we are given that $OA/AD > a_1/b_1$ (where a_1 and b_1 are both numbers). From this we deduce that $OA^2/AD^2 > a_1^2/b_1^2$, and then that

$$\frac{OD^2}{AD^2} = \frac{OA^2 + AD^2}{AD^2} > \frac{a_1^2 + b_1^2}{b_1^2},$$

and, if a_2 is the root of a perfect square less than $a_1^2 + b_1^2$,

$$\frac{OD}{AD} > \frac{a_2}{b_1}.$$

But since OE is the bisector of the angle AOD we have

$$\frac{OA}{EA} = \frac{DO}{ED} = \frac{OA + DO}{AD} > \frac{a_1 + a_2}{b_1}.$$

Therefore,

$$\frac{OA}{EA} > \frac{a_1 + a_2}{b_1} = \frac{a_3}{b_1}.$$

So we can construct the increasing series of numbers $a_1, a_2, a_3, \ldots$ such that

$$a_2 < \sqrt{a_1^2 + b_1^2}, \qquad a_3 = a_1 + a_2, \qquad a_4 < \sqrt{a_3^2 + b_1^2},$$

etc. . . . .

Archimedes starts (Fig. 7.3) with an angle of 30°, so the first value of OA/AC is $\sqrt{3}$, which is greater than $\frac{265}{153}$.

He then constructs the series

$$b_1 = 153, \qquad a_1 = 265, \qquad a_2 = 591\tfrac{1}{8} < \sqrt{a_1^2 + b_1^2},$$

etc. . . , obtaining the result that

$$\frac{OA}{AG} > \frac{4673\frac{1}{2}}{153}.$$

He notes that AG is half the side of a 96-sided polygon inscribed in the circle and the proposition is then proved.

Second part (Fig. 7.6). AD, the bisector of the angle BAC, cuts the circle in D and the line BC in d.

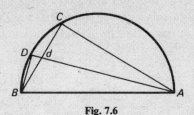

Fig. 7.6

Now $AC/CB < a_1/b_1$ (where a_1 and b_1 are given numbers), and $AB/CB < a_2/b_1$. We require an upper bound for the value of AD/AB. Since the triangles dDB and BDA are similar, we have

$$\frac{AD}{BD} = \frac{BD}{Dd} = \frac{AB}{Bd}.$$

But by the theorem relating to the bisector we have

$$\frac{AB}{Bd} = \frac{AB + AC}{CB} < \frac{a_2 + a_1}{b_1}.$$

Therefore,

$$\frac{AD}{BD} < \frac{a_2 + b_1}{b_1} = \frac{a_3}{b_1}.$$

so

$$\frac{AB^2}{BD^2} = \frac{AD^2 + DB^2}{BD^2} < \frac{a_3^2 + b_1^2}{b_1^2},$$

and if

$$a_4 > \sqrt{a_3^2 + b_1^2}, \qquad \text{we have } \frac{AB}{BD} < \frac{a_4}{b_1}.$$

We have thus constructed a series of numbers $a_1, a_2, a_3,$ a_4, by finding square roots which are larger than the number we require.

On two occasions Archimedes also simplifies the ratios he obtains, so as to avoid using very large numbers. The results of his calculations are summarized in the following table, which we have taken from Paul Tannery. The notation has been slightly modified.

	Circumscribed Polygons			Inscribed Polygons		
Number of sides	a odd sub-scripts	a even sub-scripts	b	a odd sub-scripts	b even sub-scripts	b
6	265	306		1351	1560	780
12	571	$591\frac{1}{8}$		2911	$3013\frac{1}{2}\frac{1}{4}$	780
24	$1162\frac{1}{8}$	$1172\frac{1}{8}$	153	$\begin{cases} 5924\frac{1}{2}\frac{1}{4} \\ 1823 \end{cases}$	$1838\frac{9}{11}$	$\begin{matrix} 780 \\ 240 \end{matrix}$
48	$2334\frac{1}{4}$	$2339\frac{1}{4}$		$\begin{cases} 3661\frac{9}{11} \\ 1007 \end{cases}$	$1009\frac{1}{6}$	$\begin{matrix} 240 \\ 66 \end{matrix}$
96	$4673\frac{1}{2}$			$2016\frac{1}{6}$	$2017\frac{1}{4}$	66

From this we conclude that

$$3\frac{1}{7} > \frac{96 \times 153}{4673\frac{1}{2}} > \pi > \frac{96 \times 66}{2017\frac{1}{4}} > 3\frac{10}{71}.$$

For a long time this result was used to check the results obtained by any method proposed for squaring the circle. Thus when Ptolemy adopted for π a sexagesimal value of 3;8,30, a value that can be deduced from his astronomical tables, he noted by way of justification that this value is very close to the mean of the two bounds established by Archimedes. We should nonetheless note that Ptolemy's value is the best value of π expressible in three sexagesimal places.

Archimedes's procedure for finding approximate numerical values of π (without, of course, referring to π as a number), by establishing narrower and narrower limits between which the value must lie, turned out to be the only practicable way of squaring the circle. But the Greeks also tried to square the circle exactly, that is they tried to find a method, employing only straight edge and compasses, by which one might construct a square equivalent to the given circle. All such attempts failed, though Hippocrates of Chios did succeed in squaring lunes.

Hippocrates begins by noting that the areas of similar segments of circles are proportional to the squares of the chords which subtend them.

1. Consider a semi-circle ACB with diameter AB (Fig. 7.7). Let us inscribe in this semi-circle an isosceles triangle

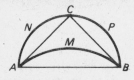

Fig. 7.7

ACB, and then draw the circular arc AMB which touches the lines CA and CB at A and B respectively. The segments ANC, CPB and AMB are similar. Their areas are therefore proportional to the squares of AC, CB and AB respectively, and from Pythagoras's Theorem the greater segment is equivalent to the sum of the other two. Therefore the lune $ACBMA$ is equivalent to the triangle ACB. It can therefore be squared.

2. By means of a straight edge and compasses we can construct an isosceles trapezium $ABCD$ such that $AB = BC = CD$ and $AD^2 = 3AB^2$ (Fig. 7.8). If we draw the

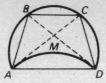

Fig. 7.8

circular arc *ABCD*, and then the arc *AMB*, to touch *AC* at
A, the segments *AB*, *BC*, *CD* and *AMD* are similar. Since
their areas are proportional to the squares of their bases
we can deduce from the previous result that the large segment is equivalent to the sum of the three others. The
lune *ABCDMA* is equivalent to the trapezium *ABCD*.

3. Consider an isosceles trapezium *ABCD* such that
$AB = BC = CD$ (Fig. 7.9).

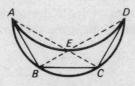

Fig. 7.9

Let the diagonals *AC* and *BD* intersect in *E*. The circle
circumscribed around the triangle *AED* touches the lines
AB and *DC*, as can be seen by considering the angles
involved. In the circle *ABCD* there are thus three equal
segments *AB*, *BC* and *CD*. The segments *AE* and *ED* of the
circle *AED* are similar to the former three segments.
If $2AE^2 = 3AB^2$ these two segments are equivalent to the
three others, and the lune *ABCDEA* is equivalent to the
concave pentagon *ABCDE*. It can therefore be squared.
If we know *AB*, we can construct *AE* by means of a straight
edge and compasses. Hippocrates says that to complete
the construction of the trapezium we should draw a circle
with centre *B* and passing through the point *C* (Fig. 7.10),
construct the perpendicular bisector of the line *BC* and then

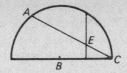

Fig. 7.10

draw a straight line, AE, of the required length, between the circle and the bisector, such that it points towards C, that is, such that it would pass through C if it were produced.

Hippocrates does not say what method other than trial and error should be used to carry out this construction, which belongs to the category of vergings. We can easily show that the problem can be solved by a construction using only a straight edge and compasses. From the definition of E we have $AE \times EC = BC^2 - BE^2$ [Euclid, Book III]. But $BE = EC$ (since E lies on the bisector), therefore $AE \times EC = BC^2 - EC^2$, i.e. $EC [AE + EC] = BC^2$ or $EC \times AC = BC^2$. We are thus required to construct two lengths CE and AC given their difference, AE, and their geometric mean BC. This is a canonical problem [Euclid, Book VI, Proposition 29, or Data, Proposition 59].

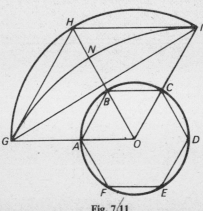

Fig. 7.11

4. Consider two concentric circles with common centre O and radii such that the square of the radius of the larger circle is six times the square of the radius of the smaller one (Fig. 7.11). Let us inscribe in the smaller circle the regular hexagon $ABCDEF$. Let OA cut the larger circle in G, the line DB in H and the line OC in I. On the line GI we construct a circular segment GNI similar to the segment GH. Hippocrates shows that the lune $GHIN$ plus the smaller circle is equivalent to the triangle GHI plus the hexagon.

We shall now consider another Greek method of squaring the circle, which is in effect a method of rectifying the circumference.

We have already mentioned the *quadratrix* of Hippias, in connection with the problem of trisecting an angle (see page 167).

It seems that in the fourth century BC Dinostrates, a pupil of Eudoxus, discovered a relation between the radius of the circle and the distance CH from the centre of the circle to the foot of the quadratrix.

Pappus of Alexandria explains as follows [2]:

PROPOSITION 26. If $ABCD$ is a square and BED the arc of a circle with centre C, while BGH is a quadratrix generated in the aforesaid manner [Fig. 7.12], it is proved that the ratio of the arc DEB towards the straight line BC is the same as that of BC towards the straight line CH. For if it is not, the ratio of the arc DEB towards the straight line BC will be the same as that of BC towards either a straight line greater than CH or a straight line less than CH.

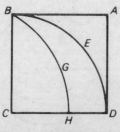

Fig. 7.12

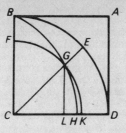

Fig. 7.13

Let it be the former, if possible, towards a greater straight line CK, [Fig. 7.13], and with centre C let the arc FGK be drawn cutting the curve at G, and let the perpendicular GL be drawn, and let CG be joined and produced to E. Since therefore the ratio of the arc DEB towards the straight line BC is the same as the ratio of BC, that is CD, towards CK, and the ratio of CD towards CK is the same as that of the arc BED towards the arc FGK (for the arcs of circles are in the same ratio as their diameters), it is clear that the arc FGK is equal to the straight line BC. And since by the property of the curve the ratio of the arc BED towards ED is the same as the ratio of BC towards GL, therefore the ratio of FGK towards the arc GK is the same as the ratio of the straight line BC towards GL. And the arc FGK was proved equal to the straight line BC; therefore the arc GK is also equal to the straight line GL, which is absurd. Therefore the ratio of the arc BED towards the straight line BC is not the same as the ratio of BC towards a straight line greater than CH.

I say that neither is it equal to the ratio of BC towards a straight line less than CH. For, if it is possible, let the ratio be towards KC, and with centre C let the arc FMK be described, and let KG at right angles to CD cut the quadratrix at G, and let CG be joined and produced to E [Fig. 7.14].

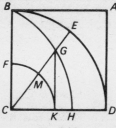

Fig. 7.14

In similar manner to what has been written above, we shall prove also that the arc FMK is equal to the straight line BC, and that the ratio of the arc *BED* towards *ED*, that is, the ratio of *FMK* towards *MK*, is the same as that of the straight line *BC* towards *GK*. From this it is clear that the arc *MK* is equal to the straight line *KG*, which is absurd. The ratio of the arc *BED* towards the straight line *BC* is therefore not the same as the ratio of *BC* towards a straight line less than *CH*. Moreover it was proved not the same as the ratio of *BC* towards a straight line greater than *CH*; therefore it is the same as the ratio of *BC* towards *CH* itself.

This also is clear, that if a straight line is taken as a third proportional to the straight lines *HC*, *CB* it will be equal to the arc *BED*, and four times this straight line will be equal to the circumference of the whole circle. A straight line equal to the circumference of the circle having been found, a square can easily be constructed equal to the circle itself. For the rectangle contained by the perimeter of the circle and the radius is double of the circle, as Archimedes demonstrated.

Archimedes describes another transcendental curve, now known as the Archimedean Spiral, which like Hippias's quadratrix can be used in connection with the problem of rectifying the circumference of a circle. In the preface to the treatise *On Spirals* Archimedes says [3]:

> ... If a straight line of which one extremity remains fixed be made to revolve at a uniform rate in a plane until it returns to the position from which it started, and if, at the same time as the straight line revolves, a point move at a uniform rate along the straight line, starting from the fixed extremity, the point will describe a spiral in the plane .... And, if a straight line touch the spiral at the extreme end of the spiral and another straight line be drawn at right angles to the line which has revolved and resumed its position from the fixed extremity of it, so as to meet the tangent, I say that the straight line so drawn to meet it is equal to the circumference of the circle.

We should note that Archimedes and Hippias merely replace the problem by an equivalent one: instead of being required to rectify the circumference of the circle we are now required to find an accurate way of determining the position of the vertex of the quadratic or of constructing the tangent to the spiral.

We shall give only a brief history of the problem of squaring the circle. The Arabs made very elaborate calculations of approximate values of π (representing it as a ratio) but in the West it was not until the fifteenth century that the problem began to attract much attention on the part of mathematicians. One of those responsible for this revival of interest was Nicolaus Cusanus, who worked on the problem for about ten years. Regiomontanus later pointed out that Cusanus had made many errors in his reasoning, and this criticism stimulated others to further work.

One of those who tried to find a theoretical solution to the problem of squaring the circle was an engineer called Simon du Chêne (or Simon Van der Eyck or Simon a Quercu), who had been born at Dôle, in France, but was now working in the Netherlands. His book about the problem, written towards the end of the sixteenth century, was dedicated to the Prince of Orange, who showed it to another engineer, Adriaan Anthonitz of Metz. (Anthonitz's two sons, who both became famous, took the surname Metius.) Adriaan Anthonitz suggested to Ludolph van Ceulen that, as a check, he should work out limits for the value of π more precise than those found by Archimedes.

It was thus that van Ceulen set about his famous calculations, using only the methods employed by Archimedes. These calculations eventually enabled him to find a value of π correct to 34 figures.

However, at his first attempt, in 1586, van Ceulen merely showed that the ratio in question was greater than the ratio 3 141 557 587 to 1 000 000 000 but less than the ratio 3 141 662 746 to the same power of ten.

Adriaan Anthonitz immediately concluded, by a method he did not explain but which must have been somewhat like Euclid's Algorithm and the method of continuous fractions, that the ratio must lie between $3\frac{15}{126}$ and $3\frac{17}{120}$. Using these two limits he found an intermediate value of

$$3\frac{15 + 17}{106 + 120} = 3\frac{32}{226} = 3\frac{16}{113} \qquad \text{i.e.} \ \frac{355}{113}.$$

This value is known as *Metius's approximation*. It was later found to be less than 10^{-6} greater than the exact value of π.*

Many other mathematicians of the late sixteenth and early seventeenth centuries also used Archimedes's methods, or some simplified version of them, to calculate good approximate values of π, though none of them took their calculations as far as van Ceulen's.

The last mathematician to use the Greek geometrical method in this way was Huygens, but he also worked in positional decimal arithmetic. His first work on the subject was published in 1651 and was entitled *Theoremata de quadratura hyperboles, ellipseos et circuli ex data portionum gravitatis centro*; then in 1654, when he was 25, he published the short treatise *De circuli magnitudine inventa*.

This latter work shows the extent of the influence the mathematics of Ancient Greece has exerted over the mathematics of modern times, and stands as an example of the continuity of scientific thought over the centuries.

This treatise by the young Huygens brings to an end the first part of the history of the squaring of the circle.

---

* In his *Practical Geometry* of 1611 Adriaan Metius writes:

Parens meus Illustrium DD. Ordinum Confoederatarum Belgiae Provinciarum Geometra, in libello quem conscripsit adversus Quadraturam circuli Simonis a Quercu demonstravit proportionem peripheriae ad suam diametrum esse minorem quam $3\frac{17}{120}$ hoc est $\frac{377}{120}$, majorem vero quam $3\frac{15}{106}$ hoc est $\frac{333}{106}$, quarum proportionum intermedia existit $3\frac{16}{113}$ sive $\frac{355}{113}$. Quae quidem intermedia proportio aliquantulum existit major, quam ea, quam invenit M Ludolph a Collen, cujus tamen differentia est minor quam $\frac{1}{1\,000\,000}$.

The last part of this passage is interesting: Adriaan Metius speaks of the difference of two ratios not in the Greek sense of the word, which we should now call the 'quotient', but in the modern sense, where the ratios are identified with numbers.

The modern, or analytical, period had already begun a few years earlier, with the work of Vieta. We cannot follow further developments in detail, but we shall summarize them briefly.

Using one of Euclid's figures, Vieta showed that the ratio of the area of a square inscribed in a circle to the area of the circle itself could be expressed as the product $\sqrt{\frac{1}{2}}\sqrt{\frac{1}{2} + \frac{1}{2}\sqrt{\frac{1}{2}}}\ldots$, whose terms are generated according to the rule that given one term, a, the following term is $\sqrt{\frac{1}{2} + a/2}$.*

John Wallis, in his *Arithmetica Infinitorum* of 1655, gave another, now famous, value of π expressed as an infinite product:

$$\frac{\pi}{2} = \frac{2}{1} \cdot \frac{2}{3} \cdot \frac{4}{3} \cdot \frac{4}{5} \cdot \frac{6}{5} \cdot \frac{6}{7} \cdot \frac{8}{7} \cdot \frac{8}{9} \ldots \text{ ad infinitum.}$$

Wallis wrote his result as follows, using □ to designate the number $4/\pi$, the ratio of the area of the circumscribed square to the area of the circle

minor quam

$$\frac{3 \times 3 \times 5 \times 5 \times 7 \times 7 \times 9 \times 9}{2 \times 4 \times 4 \times 6 \times 6 \times 8 \times 8 \times 10}$$

$$\frac{\times 11 \times 11 \times 13 \times 13}{\times 10 \times 12 \times 12 \times 14} \times \sqrt{1\frac{1}{13}}$$

□

major quam

$$\frac{3 \times 3 \times 5 \times 5 \times 7 \times 7 \times 9 \times 9}{2 \times 4 \times 4 \times 6 \times 6 \times 8 \times 8 \times 10}$$

$$\frac{\times 11 \times 11 \times 13 \times 13}{\times 10 \times 12 \times 12 \times 14} \times \sqrt{1\frac{1}{14}}.$$

---

* Vieta published this result in his *Variorum de rebus Responsorum libri VIII* of 1593.

So if we carry the calculation of $\square$ only as far as the factors

$$\frac{2n-1}{2n-2} \quad \text{and} \quad \frac{2n-1}{2n},$$

and then multiply the result by

$$\sqrt{1\frac{1}{2n-1}} = \sqrt{\frac{2n}{2n-1}},$$

we obtain an upper bound for the number (in Wallis's work ratios are identified with numbers—modern mathematical usage was becoming established), but if we multiply by

$$\sqrt{1\frac{1}{2n}} = \sqrt{\frac{2n+1}{2n}},$$

we obtain a lower bound for the number.

This interest in obtaining as close bounds as possible for the value of π can be traced back to the work of Archimedes. Using P_{2n} to designate the product as far as the terms

$$\frac{2n-1}{2n-2}, \frac{2n-1}{2n}$$

we have

$$P_{2n}\sqrt{\frac{2n+1}{2n}} < \square < P_{2n}\sqrt{\frac{2n}{2n-1}}.$$

Therefore the error is less than the difference between the two bounds, i.e. less than

$$\frac{P_{2n}}{\sqrt{2n(2n-1)}} \simeq \frac{\square}{2n}.$$

It is clear that Wallis's infinite product tends only very slowly towards its limiting value. Since contemporaries had questioned the correctness of the formula, Lord Brouncker (1620–1684), a friend of Wallis, undertook some calculations to verify it, and showed that it gave the following bounds for the ratio of the circumference of a circle to its diameter:

the ratio is greater than 3·141592653569 ... to 1

and less than 3·141592653696 ... to 1,

which is a very satisfactory result.

Brouncker also found an expression for the value of □ as an infinite continued fraction:

$$\square = 1 + \cfrac{1}{2 + \cfrac{9}{2 + \cfrac{25}{2 + \cfrac{49}{2 + \cfrac{81}{2 + \text{etc.} \ldots}}}}}$$

This means, in effect that □ is greater than the fractions

$$1, \quad 1 + \cfrac{1}{2 + \cfrac{9}{5}}, \quad 1 + \cfrac{1}{2 + \cfrac{9}{2 + \cfrac{25}{2 + \cfrac{49}{9}}}}, \quad \ldots \text{etc.}$$

but less than the fractions

$$1 + \cfrac{1}{3}, \quad 1 + \cfrac{1}{2 + \cfrac{9}{2 + \cfrac{25}{7}}}, \quad \ldots \text{etc.}$$

Lord Brouncker's formula, like Wallis's formula is interesting, but it does not provide a practical means of calculating the value of π.

A little later, James Gregory, Mercator, Newton and Leibniz found the following expression for an angle as the sum of an infinite power series in terms of its tangent:

$$\tan^{-1} x = x - \frac{x^3}{3} + \frac{x^5}{5} - \frac{x^7}{7} + \cdots,$$

and Newton found the expression for the angle in terms of its sine:

$$\sin^{-1} x = x + \frac{1}{2} \cdot \frac{x^3}{3} + \frac{1}{2} \cdot \frac{3}{4} \cdot \frac{x^5}{5} + \frac{1}{2} \cdot \frac{3}{4} \cdot \frac{5}{6} \cdot \frac{x^7}{7} + \cdots.$$

Taking $x = 1$, the first, and simpler, of these two formulae gives

$$\frac{\pi}{4} = 1 - \frac{1}{3} + \frac{1}{5} - \frac{1}{7} + \cdots.$$

This series converges too slowly to be useful for actual computation, but we can express the angle $\pi/4$ as the sum of smaller angles, using the formula

$$\tan (A + B) = \frac{\tan A + \tan B}{1 - \tan A \tan B},$$

obtaining, for example,

$$\frac{\pi}{4} = 4 \tan^{-1} \left(\frac{1}{5} \right) - \tan^{-1} \left(\frac{1}{239} \right)$$

or

$$\frac{\pi}{4} = \tan^{-1} \left(\frac{1}{2} \right) + \tan^{-1} \left(\frac{1}{3} \right).$$

Methods like these have been in use since the beginning of the eighteenth century, but before we consider their further development let us first see how it came about that the Greek letter π was used to denote the ratio of the circumference of a circle to its diameter.

On page 343 of the *Lectiones habitae in Scholis Publicis Academiae Cantabrigiensis* of Newton's teacher Isaac Barrow (1630–1667) we read:

Theor. II—Circulus aequatur dimidio rectangulo ex circumferentia et radio (Vel triangulo, cujus basis aequatur circumferentiae altitudo radio).

Hoc est, posito (ut semper post hac) circumferentiam vocari π, et radium R (vel r) et diametrum δ (vel D) $\bigcirc = r/2 = \delta\pi/4$.

It is clear that in this passage Barrow is using π to denote the length of the circumference, so what we should now call π would to Barrow have been π/D, the ratio of the circumference to the diameter. The reason for the abbreviation becomes clear if we remember that in his treatise *Measurement of a Circle* (see above, pp. 186–193). Archimedes calls the length of the circumference of the circle its 'perimeter', $\pi\varepsilon\rho\acute{\iota}\mu\varepsilon\tau\rho\sigma\varsigma$.

In 1647 Oughtred had used the letter π in the same way as Barrow.*

At the time of Barrow and Brouncker, mathematicians were beginning to identify ratios with numbers. For instance, in 1706, W. Jones represented the ratio 3·14... by the letter π. Jean Bernoulli used c. In 1747 Euler used p, although in a letter of 1736 he had used c. In 1742 Golbach used π. In his famous *Introduction to Infinitesimal Analysis* (published in Latin in 1748), which for a long time was the standard work on analysis, Euler used π, and this usage later became universal.

Various mathematicians computed values of π to a large number of decimal places. Sharp, encouraged by Halley, calculated π to 72 decimal places (71 of them correct) in 1699.

Machin (before 1706) calculated 100 decimal places (all correct).

---

* Wallis and Barrow adopted various usages from Oughtred, who gives the following expressions for the area of the circle and the volume of a cone:

$$\frac{\pi}{\delta}Rq \text{ est area circuli}: \frac{\pi}{3\delta}Rq \times \text{altitud. est conus.}$$

Lagny (1719), gave 127. (The 113th is wrong: he gave 7 instead of 8).

Vega (1789) calculated 143. (126 are correct. He corrected Lagny's error). In 1794 Vega calculated 140 (136 correct).

Callet, in his tables of 1795 gave 154 (152 correct).

Rutherford (1841) gave 208 (152 correct).

Dahse (1844) gave 205 (200 correct).

Clausen (1847) gave 250 (248 correct).

Rutherford later (1853) gave 440 (all correct).

William Shanks gave 530 in 1853, 607 later in the same year, and in 1873 gave the 707 to be seen in the Palais de le Découverte* (only 527 of these figures are correct).

Calculations began again in America in 1946, this time using powerful mechanical calculators.

Ferguson and Wrench, working independently, both found identical values of π to 808 figures.

Since then, electronic computers have made it possible to calculate π to more than 10 000 decimal places.

We have not however answered the crucial question: is it possible to square the circle using only a straight edge and compasses?

Let us begin by reminding ourselves of some definitions. An *algebraic number* is the root of an algebraic equation with integral coefficients. A *transcendental number* cannot be the root of such an equation. A *rational number* is an ordinary fraction. We can show that if it were possible to square the circle then π would be an algebraic number, though one with very special properties.

The many unsuccessful attempts at squaring the circle led some sixteenth century mathematicians, such as Stifel and Maurolico, to doubt whether the problem was in fact soluble.

The first mathematician to attempt to prove that squaring the circle was impossible was James Gregory, in his *Vera circuli et hyperbollae quadratura* of 1667. He in fact tried to

---

* The French equivalent of the Science Museum. J.V.F.

show that π was a transcendental number (though he naturally used somewhat different terminology), but his attempt, though very interesting, was not successful. Huygens made detailed and rather biased criticisms of it.

Less ambitiously, but with more success, Lambert in 1761 used the properties of continued fractions to show that π was irrational (i.e. that it could not be expressed as a fraction). Legendre used a slightly modified version of Lambert's argument to give a rather simple proof that π^2 was also irrational.

In 1851, Liouville showed, again by using continued fractions, that there was an infinite number of transcendental numbers (a result which had previously been suspected but had not been proved); and in 1872 Hermite was able to show that the number e, which plays an important part in the theory of logarithms and in analysis generally, was transcendental. In 1882, Lindemann, following Hermite's method, showed that π was also transcendental.

The problem of squaring the circle was at last solved: it had been proved to be insoluble, as James Gregory had foreseen in 1667.

Bibliographical References

Chapter 2

[1] Thureau-Dangin, *Textes Mathématiques Babyloniens*, Pb. 168, p. 82.
[2] *Ibid*, No. 207, p. 103.
[3] *Ibid*, plate XIV.
[4] *Mathematical Works of Sir Isaac Newton*, ed. and tr. D. T. Whiteside, Cambridge University Press.

Chapter 3

[1] Paul Tannery, 'Notice sur les lettres de Rhabdas'. *Mémoires*, t. IV, p. 133
[2] Thureau-Dangin, *op. cit.*, pp. 1, 2, 3. B.M. 13901.

[3] *Ibid*, p. 126

Chapter 4

[1] *Proclus: A Commentary on the first book of Euclid's Elements*, tr. Morrow. Princeton, 1970.

[2] *Ibid*.

Chapter 5

[1] From *Great Books of the Western World*, tr. R. Catesby Taliaferro. Encyclopedia Britannica.

Chapter 6

[1] From *Greek Mathematical Works*, tr. Thomas. Heinemann
[2] From *Greek Mathematics*, Vol. II, p. 341, tr. T. L. Heath: quoting Heron, 'Divisions of Figures', III, 20.
[3] Thomas, *op. cit.*
[4] *Appendix ad isagogon topicam* (before 1638). From the translation by P. Tannery, but reverting to Fermat's original notation.
[5] Tr. J. V. Field from the translation by Ver Eecke, slightly abridged.

[6] Tr. Thomas, *op. cit*. Lettering of figures has been amended.

Chapter 7

[1] From *Greek Mathematics*, tr. T. L. Heath.
[2] Tr. Thomas, *op. cit*. Lettering of figures has been amended.
[3] Tr. Heath, *op. cit*.

General Index